SMALL-SCALE MUNICIPAL SOLID WASTE ENERGY RECOVERY SYSTEMS

SMALL-SCALE MUNICIPAL SOLID WASTE ENERGY RECOVERY SYSTEMS

Gershman, Brickner & Bratton, Inc.

VAN NOSTRAND REINHOLD COMPANY
_____ New York

Library of Congress Catalog Card Number: 85-26388
ISBN: 0-442-22778-7

Manufactured in the United States of America

Published by Van Nostrand Reinhold Company Inc.
115 Fifth Avenue
New York, New York 10003

Van Nostrand Reinhold Company Limited
Molly Millars Lane
Wokingham, Berkshire RG11 2PY, England

Van Nostrand Reinhold
480 Latrobe Street
Melbourne, Victoria 3000, Australia

Macmillan of Canada
Division of Gage Publishing Limited
164 Commander Boulevard
Agincourt, Ontario M1S 3C7, Canada

15 14 13 12 11 10 9 8 7 6 5 4 3 2 1

Library of Congress Cataloging-in-Publication Data
Main entry under title:

Small scale municipal solid waste recovery systems.

Includes index.
1. Refuse as fuel. 2. Refuse and refuse disposal.
I. Gershman, Brickner & Bratton.
TD796.2.S63 1986   363.7'28   85-26388
ISBN 0-442-22778-7

Dedicated to all the hard-working local government staff, whose determination, through years of planning and many late nights, makes the projects happen.

Also dedicated to the families of the authors for their understanding and support of the many hours and hard work at GBB.

PREFACE

This volume is designed to give local government elected officials and their staff the background information they need on the state of the art in small-scale municipal waste-to-energy project development. It will, of course, be of interest to many others in the field. The small-scale segment of the municipal waste energy recovery industry has grown and changed in many ways in recent years.

With increasingly stringent environmental regulations pushing up the costs of landfilling, as well as today's higher prices for oil and natural gas, the economics of small-scale systems are attractive to smaller communities or counties which might at one time only have considered joining a multijurisdictional large-scale project. The difficulties involved in developing a project that envelops numerous governmental entities are discouraging, and a small, local project may be more readily achievable.

Gershman, Brickner & Bratton, Inc. hopes this book will be of assistance to those who are considering such a project, providing guidance and encouragement, as well as practical information on technologies, economics, energy markets, financing, environmental issues, and the pitfalls of project development.

Small-Scale Municipal Solid Waste Energy Recovery Systems drew upon reference data maintained in-house at GBB, as well as the publications cited in the text. Material for several chapters was originally developed by Gershman, Brickner & Bratton, Inc. for the Pennsylvania Department of Environmental Resources, published in the "Solid Waste-to-Energy Technical Manual," December 1982. The book was jointly developed by GBB senior staff, including Harvey Gershman, President; Robert Brickner, Senior Vice President; Timothy Bratton, Senior Vice President; Charles Peterson, Vice President; Frank Bernheisel, Senior Consultant; Eugene Aleshin, Senior Consultant; and Elizabeth Wood, Consultant. Overall editing and coordination was provided by Elizabeth Wood.

Especial acknowledgment is due to the GBB production staff for typing, revising, and producing the finished book: Mary Ellen Knuti, Production Manager; Kumari Tadavarthy, LaVerne Wade, Josefina Navarro, and Alicia Romain.

CONTENTS

Preface / vii

1. **Introduction** / 1

2. **Market Development of Energy Recovery** / 4
   Historical Overview / 4
   Factors Shaping the Market / 5
   Conventional Fossil Fuel and Electricity Costs / 5   Disposal
   Costs / 10   Landfill Practices / 11   Haul
   Distance / 12   Development Authorities / 13

3. **Solid Waste Characteristics** / 15
   Quantity / 15
   Composition / 17
   Energy Content / 19

4. **Markets** / 21
   Introduction / 21
   Potential Markets / 21
   Thermal Energy / 22   Steam / 23   Hot Water / 29   Hot
   Air / 30   Supply/Demand Considerations / 31
   Electricity / 31   Industrial Energy Market Summary / 34
   Cogeneration / 36
   Refuse-Derived Fuel and Materials Recovery / 37
   Refuse-Derived Fuel / 37   Ferrous Metals / 38
   Glass / 40   Aluminum / 41   Nonferrous Metals / 41
   Wastepaper / 42
   Market Development / 43
   Identifying a Market / 43   Commitment Stage / 45   Contract
   Stage / 48
   Conclusion / 48

5. **Small-Scale Technology** / 49
   Mass Burning / 49
   Direct Combustion / 50   Modular Incineration / 67
   Problems / 74
   Refuse-Derived Fuel (RDF) / 75
   RDF Processing / 75   Refuse-Derived Fuel Co-fired with

Coal / 77 Refuse-Derived Fuel Dedicated Boiler / 80
Background / 84 Current Status / 87 Future Developments / 88
Vendors / 88 Problems / 90

6. **Small-Scale System Economics / 93**
   Introduction / 93
   The Base Case—Conventional Waste Management Systems / 94
   Transportation Network / 97
   Existing Refuse Collection System / 98 Analysis
   Parameters / 98 Calculation of Haul Distances / 99 Residue
   Disposal / 100 Additional Considerations / 100 Transportation
   Cost Analysis / 100 Transportation System Cost Estimation
   Procedure / 101
   Waste-to-Energy Project Costs / 104
   Capital Costs / 108
   Operations and Maintenance Costs / 111
   Waste-to-Energy Project Economics / 122
   Revenues / 126
   Life Cycle Costs / 127
   Economics of Cogeneration Systems / 130

7. **Financing Alternatives for Small-Scale Solid Waste-to-Energy**
   Projects / 137
   Introduction / 137
   Types of Tax-Exempt Financing Available to Waste-to-Energy
   Facilities / 139
   Introduction / 139 General Obligation Bond
   Financing / 139 Municipal Revenue Bonds / 141 Industrial
   Development Revenue Bonds / 142 Leveraged Leasing / 147
   Creative Financing Techniques / 150
   Deferred Equity / 150 Step Financing (Deferred Principal
   Repayment) / 151 Variable Rate Financing (Floating Rate
   Securities) / 152 Variable Rate Demand Bonds / 152 Adjustable
   Rate Bonds / 153 Daily Adjustable Tax-Exempt Securities / 154
   Federal Tax Benefits of Private Ownership / 155
   Accelerated Cost Recovery System Depreciation / 159 Investment
   Tax Credits / 160 Interest Deductions / 162
   Typical Steps Necessary to Bring a Bond Issue to Market / 163
   Critical Project Elements Evaluated by the Investment Banking
   Community to Determine the Financeability of a Waste-to-Energy
   Project / 165 *Credit Backing / 166 Demonstrated Expertise of*
   Project Contractors and/or Engineers / 166 Independent Feasibility

Study / 167   Back-up Landfill or Other Facility
Capacity / 167   Revenue   Sufficiency / 167
Reserves / 168   Adequate Sources of Capital / 168   Continuous
Supply of Solid Waste / 168   Long-Term Market(s) for
Energy / 168   Insurance / 168   Demonstrated Compliance
with Laws and Regulations / 170   Favorable Rating by
Recognized Rating Agencies / 170

8. **Environmental Issues / 171**
Sources of Air Pollutants / 171
Types of Air Pollutants / 172
Particulates / 172   Nitrogen Oxides (NO$_x$) / 172   Sulfur Oxide
(SO$_x$) / 173   Carbon Monoxide (CO) / 173   Total Hydrocarbons
(THC) / 173   Lead (Pb) / 173   Hydrogen Chloride (HCl) and
Hydrogen Fluoride (HF) / 173   Beryllium (Be) and Mercury
(Hg) / 174   Dioxins / 174
Federal Air Pollution Control Laws and Regulations / 175
National Ambient Air Quality Standards (NAAQS) / 175
Prevention of Significant Deterioration (PSD) / 177   New Source
Performance Standards (NSPS)/ 178   National Emission Standards
for Hazardous Air Pollutants (NESHAP) / 178
Air Pollution Control Equipment / 178
Electrostatic Precipitators / 178   Fabric Filters / 179   Electrostatic
Granular Filter / 179   Scrubbers / 180
Sources and Control of Wastewater Discharge / 181
Federal Laws and Regulations / 183
National Pollutant Discharge Elimination System
(NPDES) / 183   National Pretreatment Standards / 184   Residue
and Ash / 186   Resource Conservation and Recovery Act / 187
State Environmental Standards / 189
Summary / 190

9. **Project Development / 192**
Past Examples / 192
Stages of a Project / 193
Project Building Blocks / 196
Institutional Setting / 199
Roles That Can Be Taken / 201
Procurement Alternatives and Risks / 203
Procurement Documentation / 204
Negotiations / 215
Public Involvement / 215

10. **Three Small-Scale Waste-to-Energy System Case Studies / 217**
Auburn, Maine / 217
Introduction / 217 Background / 218 Technology / 221
Markets / 223 Economics / 224 Project Evaluation / 229
Madison, Wisconsin / 232
Introduction / 232 Background / 232 Technology / 234
Markets / 236 Economics / 236 1984 Project
Summary / 238 Project Evaluation / 240
Pittsfield, Massachusetts / 241
Introduction / 241 Background / 241 Markets / 242
Technology / 246 Economics / 249 Project Evaluation / 253

Bibliography / 259

Index / 267

SMALL-SCALE MUNICIPAL SOLID WASTE ENERGY RECOVERY SYSTEMS

1
INTRODUCTION

Local governments are out on the front lines coping with massive changes in provision of public services in the United States. Technological innovations, environmental concerns, and the changing role of the federal government strain the capacity of counties, cities, towns, and villages to continue to supply the high level of service that residents have come to expect. Meanwhile, local governments operate on budgets cut sharply by the reduction of such federal programs as revenue sharing or by state or local tax limitation initiatives. However, one basic public service which must be performed, despite rising costs, is the collection and disposal of refuse.

This volume is designed to provide background and planning assistance to communities considering small-scale [50 to 500 TPD (tons per day)] municipal waste-to-energy projects.

For most communities, the initial impetus toward such an undertaking is a "garbage crisis"—the current landfill capacity will be filled in the next few years; more stringent environmental regulations may either force the landfill to close or require costly containment measures. Often, there is public opposition to siting new landfills within a reasonable haul distance, while the cost of long-distance hauling is prohibitive.

A waste-to-energy (or "resource recovery") project does not automatically solve all waste disposal problems. Ash must still be landfilled along with unprocessible wastes, such as demolition and land clearing debris, and siting may also be a problem. However, the income to the project from the sale of energy will often reduce the cost to the community (paid in the form of tipping fees) for waste disposal over the life of the facility.

A waste-to-energy project also offers a stable supply of energy, priced generally below prevailing traditional energy sources. Such an asset can be used as a community development inducement, attracting new industries or assisting the expansion of existing ones. Some communities have used waste-to-energy plants for district heating and cooling systems, themselves downtown revitalization or community development tools. Projects producing electricity generally sell the output to local utilities.

Historically, because of the availability of cheap energy sources (oil and natural gas) and the open space needed for landfills, resource recovery had not developed in the United States in the 1970s to the extent that it had in

Table 1-1. Small-Scale Waste-to-Energy Systems in the United States.

YEAR OF START-UP	LOCATION	DESIGN CAPACITY (TPD)	TYPE OF TECHNOLOGY
1967	Norfolk, VA	360	Waterwall
1971	Braintree, MA	250	Waterwall
1975	Siloam Springs, AR	21	Modular
1975	Blytheville, AR	50	Modular
1975[a]	Ames, IA	200	RDF
1976	Groveton, NH	24	Modular
1977	North Little Rock, AR	100	Modular
1977	E. Bridgewater, MA	300	RDF
1977	Portsmouth, VA	160	Waterwall
1978[a]	Lane Co., OR	500	RDF
1979	Salem, VA	100	Modular
1979	Duluth, MN	400	RDF
1979[a]	Madison, WI	400	RDF
1979[a]	Tacoma, WA	500	RDF
1979	Waukesha, WI	175	Refractory
1980	Crossville, TN	60	Modular
1980	Dyersburg, TN	100	Modular
1980	Durham, NH	108	Modular
1980	Genesee Township, MI	100	Modular
1980	Lewisburg, TN	60	Modular
1980	Osceola, AR	50	Modular
1980	Hampton, VA	200	Waterwall
1980	Palestine, TX	28	Modular
1981	Auburn, ME	200	Modular
1981	Newport News, VA	40	Modular
1981	Cassia Co., ID	50	Modular
1981	Johnsonville, SC	50	Modular
1981	Batesville, AR	50	Modular
1981	Pittsfield, MA	240	Modular
1981[a]	Collegeville, MN	64	Modular
1981	Gatesville, TX	7	Modular
1982[a]	Sumner Co., TN	200	Waterwall
1982[a]	Henrico Co., VA	200	RDF
1982[a]	Windham, CT	108	Modular
1982	Waxahachie, TX	50	Modular
1982	Portsmouth, NH	200	Modular
1982	Miami, OK	108	Modular
1982	Red Wing, MN	72	Modular
1982	Park Co., MT	72	Modular
1983	Cuba (Cattaraugus Co.), NY	112	Modular
1983	Huntsville, AL	50	Modular
1983	Glen Cove, NY	250	Refractory
1983	Lakeland, FL	300	RDF
1983	Harrisonburg, VA	120	Refractory
1984	New Hanover Co., NC	200	Waterwall
1984	Pascagoula, MS	150	Modular

Table 1-1. (continued)

YEAR OF START-UP	LOCATION	DESIGN CAPACITY (TPD)	TYPE OF TECHNOLOGY
1984	Tuscaloosa, AL	300	Modular
1984	Susanville, CA	100	Refractory
1984	Oswego Co., NY	200	Modular
1984	Oneida Co., NY	200	Modular
1984	Ft. Dix, NJ	80	Modular

a Shut down.

Europe and elsewhere. When the OPEC oil embargo pointed up how dependent the country had become on supplementing domestic supplies with suddenly expensive imported fuels, alternative energy technologies received new attention. In the 1970s, federal programs concentrated on developing experimental waste-to-energy systems such as refuse-derived fuel (RDF) and pyrolysis (which produces a flammable liquid similar to fuel oils). The technical problems encountered by many of these test facilities caused a trend toward European-style mass burning systems to provide reliable waste disposal and energy production.

At the same time, the general thinking was that large-scale (500–2000 TPD) systems provided essential economies of scale. Such large projects often required the involvement of numerous jurisdictions in order to produce the required daily waste tonnage. The political difficulties of interjurisdictional cooperation have been as much of a stumbling block to project development as the technical problems of early experimental facilities.

Small-scale systems can handle the waste of a single jurisdiction, subjurisdiction, or a few nearby communities, and thus have encountered fewer political problems that can delay or frustrate project development. The development of modular, mass burn waste-to-energy incinerators, which simplify installation and thus reduce construction costs, has provided another support for small-scale development. As shown in Table 1-1, over 40 small-scale systems have been built in the United States since 1967, with the majority of these projects utilizing modular equipment. Chapter 2, which follows, traces the development of the market for resource recovery projects in greater detail.

Chapters 3 through 8 provide the reader with detailed discussions of the aspects of resource recovery critical for project development—waste supply, energy markets, technologies, economics, financing, and environmental issues. Chapter 9, "Project Development," outlines the steps involved in bringing a project from concept to construction, as well as the roles that must be played by the actors in the process. The final chapter, Chapter 10, offers three case studies of operating projects that utilize different technologies.

3

2
MARKET DEVELOPMENT
OF ENERGY RECOVERY

HISTORICAL OVERVIEW

Incineration of solid waste as an organized practice began in 1885 in the United States. From that time until the early 1970s, reduction of the volume of waste for disposal was the primary focus of incineration. Incineration typically was used in major urban areas in which large quantities of refuse were discarded and there was a shortage of conveniently located disposal sites. Energy recovery has been practiced almost from the beginning of incineration; the first such facility in the United States began operation in 1898 in New York City. The availability of relatively low-cost conventional fuels, however, generally made energy recovery non–cost effective.

In the early 1970s, a combination of factors began to change the prospects for the use of municipal solid waste (MSW) as a fuel. The most dramatic event was the rise in imported petroleum prices in late 1973. Oil prices have risen rapidly since 1973. Even though the current trend of oil prices is stable because of the temporary excess of petroleum in the world market, there is great political uncertainty associated with near term trend changes. The cost of natural gas has also risen and is anticipated to continue to rise through 1995. This increase in price for conventional fuels and electricity, along with continued uncertainty about the reliability of supply, has led energy consumers to seek alternative energy resources such as MSW.

While energy costs were climbing, solid waste disposal costs also were growing. One influence was the public demand for environmentally sound landfill operations. Pressure to locate disposal sites in areas remote from population centers contributed to the expense of landfill disposal by incurring additional charges for transporting waste. Since waste-to-energy facilities typically are located near industrial sites, close to population centers, haul costs to these facilities are often lower than to distant landfills. This is especially true for smaller systems that rely on local waste sources. Thus landfill costs, continuing to rise steadily, have improved the cost-competitive position of waste-to-energy facilities.

A number of development authorities, on both the state and the regional levels, were founded in the 1970s. These authorities were designed to assist in project development by taking advantage of the economic incentives created by increasing energy and disposal costs. Among the services typically offered are administrative, areawide coordination and management guidance, as well as financial assistance through their ability to issue tax-exempt bonds.

Since 1970, the market for waste-to-energy systems has changed significantly. One of the most important changes has been the development of the small-scale [100 to 500 TPD (tons per day)] segment of the market. Due primarily to the combination of energy and landfill price increases and the activity of development authorities, 43 small-scale energy recovery systems that use unprocessed MSW were in operation at the end of 1984.

FACTORS SHAPING THE MARKET

Conventional Fossil Fuel and Electricity Costs

As the cost of conventional fossil fuel, primarily petroleum, has increased, interest in alternative fuels such as solid waste also has grown. Once solid waste is converted into an end use energy product, such as process steam, electricity, hot water, or hot air, the characteristics of this product are identical regardless of the type of fuel used. Therefore, an important determinant of the viability of MSW as a fuel is the cost that a company or institution must pay for the energy product.

In the case of a project processing MSW into an energy product, a municipality or a waste service company typically will operate the recovery facility and sell energy to a company or institution. Thus, the revenue received by the facility operator will depend on the type of fuel displaced. Contracts for the purchase of energy usually include a discount from the cost to generate energy with conventional fuel as an inducement for the market to switch to solid waste rather than continue to use fossil fuel.

Another important factor in any decision with regard to the use of MSW as a fuel is the future cost of the displaced energy, in addition to the current cost of conventional energy. If the cost of the displaced energy rises at a faster rate than the cost to operate a facility, the net cost will become progressively less and will possibly become a profit. No waste-to-energy systems have reported a profit yet, but several are close. This scenario places energy recovery in a favorable position vis-à-vis landfill disposal. Even so, in the initial two to three years of operation, an energy recovery system usually will incur a higher unit cost than landfill (see p. 11 for a discussion of landfill costs).

Fuel oil is the primary fuel that has been displaced by waste-to-energy systems. The higher cost of imported oil and unregulated domestic petroleum compared with natural gas and coal is the reason for the emphasis on displacement of oil. As natural gas prices are decontrolled, more attention is focused on displacing this fuel. Final decontrol of most domestic natural gas prices is set to occur in 1985. Coal remains a comparatively low-cost energy resource. Wider use of coal could affect the revenues for solid waste because income is tied to savings from the displacement of the conventional fuel. Displacement of coal would result in a lower price differential and, thus, decreased revenues for a waste-to-energy system. The extent of the impact will depend on the source of the coal. Coal obtained from a new production facility will have higher costs than that from an existing mine, which would lessen the price differential.

The price of electricity also has increased as the cost of the fuels used to generate electricity has risen. Interest in the generation of electricity, especially cogeneration of electricity and steam, by industrial, commercial, and institutional facilities has been spurred by the escalation in the value of electricity. A more important incentive for the development of cogeneration was the enactment of the Public Utility Regulatory Policies Act (PURPA) in 1978. PURPA mandates that a utility purchase excess generated power at full avoided cost, which is the incremental cost to generate electricity or to purchase electricity from another utility. Included in the incremental cost are a payment for the portion of generating capacity provided by the cogeneration system plus the average fuel cost. The use of MSW for cogeneration rather than a conventional fossil fuel is based on the same economic justifications previously presented.

Fuel Oil.   The price of fuel oil, both distillate and residual, was stable from 1965 to late 1973. During that time, the annual price increases for residual and distillate oils were about $0.07 and $0.06 per million Btu, respectively. Following the Organization of Petroleum Exporting Countries' (OPEC's) oil embargo, the annual cost increase rose to an average of $0.28 per million Btu during the period 1973 to 1978. World oil prices increased more than 50 percent per year during this period. However, price controls on domestic oil wells that had been producing prior to the embargo helped to moderate the impact of OPEC price increases in the United States. A similar trend for distillate oil is shown in Table 2-1.

The combination of phased decontrol of domestic oil prices and cutoff of Iranian oil resulted in another increase in the cost of fuel oils in 1979. On an annual basis, the price of residual oil rose by $0.22 per million Btu between 1978 and 1983. By 1983 the price of residual oil was $4.12 per million Btu, which is an increase of 250 percent, or $2.49 per million Btu, since 1973. Again, similar price increases occurred with distillate oil. The significant growth in fuel oil costs demonstrates the reason that consumers of this

Table 2-1. Fuel Oil Prices For Industry, 1973–1995.[a]

	RESIDUAL OIL		DISTILLATE OIL	
	PRICE[b,c]	AVERAGE ANNUAL PRICE CHANGE	PRICE[b,c]	AVERAGE ANNUAL PRICE CHANGE
Year				
1973	1.63	—	1.89	—
1978	3.01	+ 0.28	4.04	+ 0.43
1983	4.12	+ 0.22	6.38	+ 0.47
1985	3.78	− 0.17	5.80	− 0.29
1990	5.18	+ 0.28	7.77	+ 0.39
1995	7.08	+ 0.38	10.58	+ 0.56

[a] Annual Energy Outlook 1983, Energy Information Administration, May 1984.
[b] Based on an energy content for residual oil of 6.287×10^6 Btu/bbl and 5.825×10^6 Btu/bbl for distillate oil.
[c] 1983 dollars per million Btu.

energy resource are seeking alternative fuels. Aside from the issue of price, consumers are interested in obtaining a fuel whose supply is reliable and for which there is no danger of a future embargo, such as occurred with oil in 1973 and 1979.

The price of fuel oil is forecast by the U.S. Department of Energy (DOE) to continue to increase to the year 1995.[1] The rate of growth, however, will increase somewhat above that which occurred between 1978 and 1983. For the period 1983 to 1985, residual oil prices, for example, continued to decline at an annual rate of $0.17 per million Btu, while growth rates of $0.28 and $0.38 per million Btu are predicted for 1985–1990 and 1990–1995, respectively. Should the DOE forecast prove to be valid, then residual oil prices will be approximately $7.08 per million Btu by 1995. This would be an increase of $2.96 per million Btu over 1983 prices.

The DOE forecast is based on the persistence of relative stability in the Middle East. Given the continuation of the Iran-Iraq war, the assassination of Anwar Sadat, and the general hostility among countries in the region, long-term stability appears tenuous. Therefore, the forecast prices of $7.08 per million Btu for residual oil and $10.58 per million Btu for distillate oil in 1990 seem to be optimistic predictions.

Natural Gas. Natural gas prices for industrial customers were relatively stable, as were fuel oil prices, between 1965 and 1973. The cost for industrial customers rose at an annual rate of $0.02 per million Btu during this period.

---

1. Though the price of crude oil has declined steadily since 1981, this trend is projected to halt in the near term.

Following the rise of petroleum costs after OPEC's price increase, natural gas prices also began to climb, although at a lesser rate. Even so, the average annual incremental growth in price, $0.28 per million Btu, was less than that incurred for residual oil as a result of price controls. The price of natural gas in 1978 had risen to $2.41 per million Btu, which was 20 percent less than the cost of residual oil; see Table 2–2.

To correct the growing discrepancy between the price of natural gas and fuel oil due to controls in the price of gas, the Natural Gas Policy Act (NGPA) was enacted in 1978. This law requires the gradual phaseout of price controls on natural gas at the wellhead. Under a rather complex set of rules, phaseout of price controls for most gas production will occur in 1985.

The short-term impact of NGPA in combination with other factors was an annual increase of .35% in price per million Btu between 1978 and 1983. This increase occurred because of the growth in prices for residual and distillate fuels. By 1983, the price of natural gas was $4.18 per million Btu.

The full impact of NGPA will occur after 1985. As natural gas prices are decontrolled, it generally is forecast that the cost will become equivalent to that of residual oil. This is shown by comparing prices in Tables 2–1 and 2–2.

Once natural gas and residual oil prices reach parity, the cost for each is forecast to increase at about the same rate. By 1995, the prices of natural gas and residual oil are predicted to be $8.64 and $7.08 per million Btu, respectively.

Coal.   The price of coal was stable between 1965 and 1973, as were the prices of fuel oil and natural gas. During this eight year period, the average annual increase in the price of coal was $0.02 per million Btu. Coal prices also increased following the OPEC embargo in 1973. Even so, the average annual rate of growth was only $0.05 per million Btu, which increased be-

Table 2–2. Natural Gas Prices for Industry, 1973–1995.[a]

YEAR	PRICE[b,c]	AVERAGE ANNUAL PRICE CHANGE
1973	0.99	—
1978	2.41	+ 0.28
1983	4.18	+ 0.35
1985	4.33	+ 0.08
1990	5.56	+ 0.25
1995	8.64	+ 0.62

[a] Annual Energy Outlook 1983, Energy Information Administration, May 1984.
[b] Based on an energy content of 1,026 Btu/cw ft.
[c] 1983 dollars per million Btu.

tween 1978 and 1983 to $0.21 per million Btu. In net terms, the cost of coal had risen to $1.88 per million Btu in 1983 (Table 2-3).

Forecast average annual price increases for coal between 1983 and 1995 are about equivalent to those experienced during 1973. Thus, the price per million Btu for coal will be substantially less than either oil or natural gas. At the forecast price of $2.52 per million Btu in 1995, coal would be about $5.34 per million Btu (68 percent) less expensive than the average price for residual oil and natural gas. The coal price fails to account for the cost of transporting it to the energy user and thus overstates the price differential between this fuel and residual oil or natural gas.

As the disparity between coal and natural gas or fuel oil grows, it is anticipated that there will be substitution of coal for these two other energy resources. Such an increase in demand for coal will result in the gradual increase in the cost of coal as illustrated in Table 2-3.

Electricity.   Between 1965 and 1973 the average cost of electricity was relatively stable. During this period, the average annual increase was $0.12 per million Btu. As the rate of price increases for the fossil fuels used to generate electricity began to accelerate after late 1973, electricity costs also grew rapidly (Table 2-4). By 1978, average electricity charges to industry had risen to $10.94 per million Btu which represents an average annual growth of $0.78 per million Btu during the previous five year period.

The forecast for electricity prices by DOE indicates a continuation in the rapid rise in cost to industry. Prices of $16.62 and $17.15 per million Btu are predicted for 1990 and 1995, respectively. These increases will keep electricity the most expensive energy resource consumed by industry. However, the use of electricity by industry will continue to rise because the overall efficiency of electric processes compensates for the higher cost per Btu. In addi-

Table 2-3. Steam Coal Prices for Industry, 1973-1995.[a]

YEAR	PRICE[b,c]	AVERAGE ANNUAL PRICE CHANGE
1973	0.94	—
1978	1.97	+0.21
1983	1.88	−0.02
1985	1.98	+0.05
1990	2.17	+0.04
1995	2.52	+0.07

[a]Annual Energy Outlook 1983, Energy Information Administration, May 1984.
[b]Based on an energy content of 22.65×10^6 Btu/short ton.
[c]1983 dollars per million Btu.

Table 2-4. Electricity Prices for Industry, 1973–1995.[a]

YEAR	PRICE[b,c]	AVERAGE ANNUAL PRICE CHANGE
1973	7.05	—
1978	10.94	+0.78
1983	16.20	+1.05
1985	16.46	+0.13
1990	16.62	+0.03
1995	17.15	+0.11

[a] Annual Energy Outlook 1983, Energy Information Administration, May 1984.
[b] Based on an energy content of 3,412 Btu/kWh.
[c] 1983 dollars per million Btu.

tion, the use of electricity provides inherent environmental benefits to the user.

Since PURPA, many utilities have negotiated long-term contracts for various types of cogeneration projects. Some utilities have provided great flexibility in structuring pricing formulas, allowing higher rates during early years than later years as long as the life cycle price has a net present value equal to or less than actual avoided costs. This flexibility makes a great difference in project economics. Utilities find themselves in widely varied state regulatory and reserve capacity settings. Some state regulatory commissions aggressively encourage PURPA, while others are laissez-faire and let project proponents press for PURPA enforcement. A project will find utilities in different capacity situations also, which greatly affect both their interest in, and their value for, avoided cost-valued power. Utilities in excess capacity situations have little interest in, and usually a low value for, purchasing new capacity from a qualifying facility under PURPA. In other cases, utilities that counted on capacity from canceled nuclear power plants and have now turned to expensive power production from either coal plants not yet built or diesel/gas peakers should be very interested in reliable small power producers. Such utilities are generally more flexible in price structure as well as able to offer an attractive price for power.

Disposal Costs

Disposal costs have escalated during the past decade for several reasons. Public demand for environmentally sound disposal practices and the need to locate new landfills in areas distant from population centers are two important factors that have contributed to the cost increase.

Landfill Practices

A common practice in many areas of the country, until recently, was open dumping. After establishment of the U.S. Environmental Protection Agency (EPA) in 1970, one of the initial solid waste projects was Mission 5000: the voluntary closing of 5,000 open dumps. Although partially successful, EPA was restricted from achieving its goal by the voluntary nature of the program.

Public pressure forced most communities to switch to sanitary landfill as the standard method of disposal. Landfill costs are higher because of the need for equipment and personnel to compress and bury the refuse received on a daily basis. Average nationwide landfill costs for a small-scale operation were about $11 per ton in 1981, whereas an open dump would have incurred an expense of $1 to $2 per ton. While a landfill improves the aesthetic and environmental quality of disposal site, other environmental problems, such as groundwater contamination, can occur at an improperly operated site.

In an attempt to remedy the environmental threat of MSW disposal sites, the Resource Conservation and Recovery Act (RCRA) was signed into law in 1976. This law gave EPA the authority to close open dumps and to upgrade the quality of landfills. The regulations derived from RCRA for nonhazardous waste disposal sites require landfill operators to control and monitor methane gas generation and surface water runoff as well as groundwater contamination by leachate. An indication of the anticipated impact of the RCRA requirements on nonhazardous landfill costs is given in Table 2-5. These costs are for three landfill sizes: 10, 100, and 300 TPD. The increase in operating costs ranges from $1.96 to $7.35 per ton for 300 and 10 TPD landfills, respectively.

The actual impact of the RCRA regulations on nonhazardous landfills is uncertain at present. The EPA has shifted the staff that was involved in

Table 2-5. Impact of RCRA on the Operating Costs of Municipal Solid Waste Landfills.[a]

	1978 DOLLARS PER TON		
	10 TPD	100 TPD	300 TPD
Pre-RCRA Costs	11.15	6.65	3.95
Post-RCRA Costs	18.50	9.73	5.91
Differential	7.35	3.08	1.96
Percentage Increase	66%	46%	50%

[a]Source: "Draft Environmental Impact Statement on the Proposed Guidelines for the Landfill Disposal of Solid Waste," U.S. Environmental Protection Agency, Washington, D.C., 1979.

nonhazardous waste activities to the hazardous waste program. In addition, direct funding of state solid waste programs by EPA was curtailed during 1980 to 1984, which means there are less staff at the state level to ensure compliance with the RCRA mandate. However, recent legislative initiatives suggest a change which will cause increased implementation of stringent landfill regulations.

Pressure by those who live near landfills probably will continue and thus force most communities to maintain relatively high standards. In addition, some states, particularly those that are more populous, have more stringent enforcement of environmental regulations. Landfill costs, therefore, will continue to rise in response to local environmental pressure and needs. The expense for labor, fuel, and equipment also will contribute to steady growth in disposal costs.

Recent calculations provide a better estimate for predicted increases in disposal costs. Table 2-6 displays 1983 costs for disposing 10 TPD through 200 TPD of solid waste. These estimates show that costs are already exceeding those predicted by EPA in 1978.

Many factors are involved in disposal of solid waste so that accurate cost forecasts are not possible on a generalized basis. Such factors include, but are not limited to, land cost, transportation, distance, geologic and hydrologic factors, population, social pressures, and regulations. For example, a significant step increase often occurs when a sanitary landfill comes under more stringent environmental regulations. Thus, one normally can only provide estimated ranges in costs.

Haul Distance

As already alluded to, various parameters have combined to force landfill operations to move to areas remote from population centers as existing sites close. Citizen opposition to new landfills and the high cost of land (with many competing uses) located near population centers are two important factors that have led to the trend to distant landfill sites.

Location of sites away from the point of generation has increased travel time, which means a corresponding rise in the cost for labor and equipment to transport wastes. Use of transfer hauling can lower haul costs compared

Table 2-6. Sanitary Landfill Disposal Costs.

1983 DOLLARS PER TON				
10 TPD	25 TPD	50 TPD	100 TPD	200 TPD
42.50	30.00	22.20	21.20	15.90

with direct haul in collection trucks. However, transfer hauling operations are expensive. For example, the transfer haul cost can be in the $10–15 per ton range to get a landfill 25 miles away. Without transfer haul, the transportation cost would have been much higher.

Energy recovery systems can result in lower average transportation costs. If the energy market is more centrally located than a disposal site, substantial hauling costs can be avoided. Haul time to such a facility would be less than to a distant landfill site for most participating communities. The average haul cost for the system, therefore, would be lower. If an average haul cost charge is assessed by the communities that participate in a project, then all would benefit. Furthermore, future cost increases would be less.

Development Authorities

The economic incentives provided by the growth in conventional fossil fuel prices and landfill disposal costs have provided impetus for the development of several energy recovery projects. Even so, these economic incentives alone have often been insufficient for project completion. Institutional obstacles that arise when combining waste from several communities can provide more significant barriers to successful completion of a project than technical or economic factors. To meet these institutional problems, areawide development authorities have been instituted at the state and regional levels in several areas. These authorities provide organizational, administrative, and financial assistance to prospective communities. In addition, they are able to coordinate project planning and development over a broader geographic area.

States with energy recovery development authorities are:

- Connecticut
- Delaware
- Louisiana
- Maryland
- New Jersey
- Rhode Island
- West Virginia

The Connecticut Resources Recovery Authority (CRRA), which was mandated by the state legislature in 1973, and the Northeast Maryland Waste Disposal Authority are examples of statewide and regional authorities, respectively. Each is involved in the development of small-scale energy recovery projects.

A variation of the energy recovery development authority is the environmental development agency. These agencies are enabled to finance, own, and operate environmentally related facilities such as waste-to-energy plants. Among the states with this type of agency are Maryland (Maryland Environmental Service) and New York (New York State Environmental Facilities Corporation).

3
SOLID WASTE CHARACTERISTICS

The energy that can be recovered from solid waste is limited by the characteristics of the refuse discarded by the community or group of communities served by a recovery system. Quantity, composition, and energy content are the three important characteristics covered in this chapter.

QUANTITY

Municipal sold waste (MSW) is comprised of discards from residences, institutions, and commercial establishments, as well as nonhazardous, light industrial refuse (e.g., corrugated containers). Residential waste quantities are determined by population and generation rate (pounds per capita per day). The generation rate will vary based on the level of disposable income available to a community to spend on goods and services. The by-products of this consumption become the waste discarded by householders.

Waste quantities from commercial and light industrial establishments are affected by the type of operation, such as grocery stores, department stores, and warehouses, and the generation rate (pounds per employee per day). As with residences, the generation rate is influenced by the level of economic activity. A period of high sales results in greater waste generation than a slow period. The size of institutions, such as schools/colleges, hospitals, and prisons, combined with the related generation rate, determines the waste quantities discarded from this source.

The degree to which residences, institutions, and business establishments separate materials for recycling also will influence the quantity of waste discarded. Separation of corrugated is standard practice in many commercial stores. In the last ten years, materials recycling (source separation) by householders has become more common. Many cities and community groups sponsor recycling centers or separate collection programs, in which recyclables, usually newspaper, are collected from the curbside.

Nationwide estimates of the rate at which MSW is generated have been developed by public and private organizations. These estimates are simply average generation rates, which fail to account for local variations in income level and the types of businesses and institutions in a particular community.

Caution, therefore, should be taken in applying these figures to a specific area unless one is simply developing an order-of-magnitude estimate of the quantity of waste being discarded. With these average generation rate and population data, a rough calculation can be made of the quantity of residential, institutional, and commercial and light industrial solid waste discarded in an area. An improved approach is to use generation rate data developed specifically for another community, with similar characteristics, in the same state or region. The best method to determine waste quantity is to have actual weigh data.

A current (1981) estimate of municipal waste discards nationwide of 3.48 pounds per capita per day was developed from U.S. Environmental Protection Agency (EPA) data. At this rate, a community or group of communities with a population in the range of 46.0 to 229.9 thousand would have had in 1981 a daily waste load that was compatible with a waste-to-energy system with a design capacity of 100 to 500 TPD (tons per day), Table 3-1. This is the waste quantity range suitable for small-scale energy recovery, based on an average annual throughput of 80 percent of design capacity. A slightly lower population will be needed to generate this range of solid waste in the future, due to growth in the waste generation rate. Based on anticipated economic growth, the generation rates for 1985 and 1990 were forecast to be 3.50 and 3.51 pounds per capita per day, respectively.

Towns with less than 100 TPD of MSW have developed waste-to-energy systems. However, most such systems have used the waste from a local industry to supplement the municipal discards. Industrial process waste, such as sawdust or plastic trimmings, has a much higher energy content per unit weight than MSW; thus a smaller quantity of material is needed to produce

Table 3-1. Approximate Population Necessary to Supply Waste-to-Energy Facilities in the 100 to 500 TPD Range with Municipal Solid Waste, 1981-1990.[a,b]

WASTE QUANTITY (TPD)	POPULATION BY YEAR (1,000)		
	1981	1985	1990
100	46.0	45.7	45.6
200	91.9	91.4	91.2
300	137.9	137.1	136.7
400	183.9	182.9	182.3
500	229.9	228.6	227.9

[a] Waste generation rates in 1981, 1985, and 1990 are 3.48, 3.50, and 3.51 pounds per capita per day, respectively.
[b] Based on an assumed average daily throughput of 80 percent of design capacity.

the same quantity of energy. Another important factor in the development of smaller plants has been very high landfill costs, which enable such small systems using MSW to be economically viable. However, such systems are very much dependent on site-specific factors.

Within the United States, there were approximately 570 communities, or groups of communities, that were estimated to have generated sufficient MSW to supply waste-to-energy facilities in the 100 to 500 TPD range during 1981. These data only indicate the maximum potential market for small-scale waste-to-energy systems based on the available waste supply. Local energy consumption patterns and demand, as well as the cost and availability of landfill disposal sites, will determine the economic viability of energy recovery in these communities.

The quantity of solid waste discarded in a given area will vary by the month and season of the year. This variation is due to changes in waste composition. For example, increased quantities of yard waste and beverage containers are discarded in the warm summer months. Seasonal changes in the quantity of waste discarded are related primarily to residential sources. There is only slight variation in the generation of commercial and industrial wastes, given a constant level of economic activity. The impact of seasonal variation on daily waste quantities, consequently, is dependent on the percentage of residential refuse to the total waste stream. On a national basis, residential waste accounts for 67 percent of MSW, while the remaining 33 percent is attributed to commercial sources.

The degree of seasonal variation will fluctuate by region of the country. In general, those areas that have a stable climate tend to have less seasonal variation than regions in which there are climatic changes. Even so, for the majority of the communities in the continental United States, solid waste will vary from about 20 percent above the average during spring cleanup (April) and late summer (August) to 20 percent below the average in mid-winter (January), as shown in Figure 3–1.

COMPOSITION

As with the quantity of waste discarded in similar size communities, the composition of this refuse also will vary. Differences in residential waste composition are due to variations in the types of goods and services purchased. States such as Oregon and Michigan that have a beverage container deposit law, for example, will have a lower percentage of glass and metal in their municipal waste than states without such legislation. The composition of commercial/industrial waste also will differ based on the types of businesses located in a community.

The composition of MSW is community specific. However, nationwide

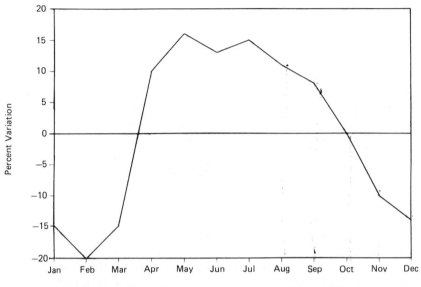

Figure 3-1. Monthly variation in solid waste quantity.

composition data that have been developed for the EPA provide an indication of the approximate percentage of the constituents in MSW (Table 3-2). No dramatic shifts in waste composition are forecast by the data in Table 3-2. Several changes will occur that have implications for energy recovery. One area of importance is the decline in inorganic, noncombustible materials, including glass and ferrous metals. The market demand for glass will continue to decrease as producers seek substitutes (e.g., plastics) which are lighter in weight and safer to use. Container and packaging applications are the areas in which the reduction in glass will be the greatest. Ferrous use also will decline in containers and packaging. This loss will occur primarily in the beverage market as aluminum containers displace bimetal cans. Aluminum's growth is forecast to continue for several years until saturation of the beverage market is reached. Stabilization is anticipated in the future as the general population matures and the demand for soft drinks and beer levels off. Plastics are forecast to grow rapidly, even though plastic is a petroleum-based product. Lightweight plastics have a total energy balance (e.g., raw materials, process energy, transportation) advantage over most other materials.

The composition of MSW will vary by the month as well as by the season of the year. The percentage of yard waste and beverage containers, as well as other convenience items such as paper plates, is higher in the summer months than in winter. This shift in waste composition by month and season is the

Table 3-2. Average Composition of MSW Nationwide, 1981-1990.[a]

TYPE OF MATERIAL	COMPOSITION (%)		
	1981	1985	1990
Glass	9.7	9.5	9.3
Metal	8.8	8.7	8.5
(ferrous)	(7.5)	(7.1)	(7.0)
(aluminum)	(1.0)	(1.2)	(1.2)
(other nonferrous)	(0.3)	(0.4)	(0.3)
Paper and Paper Products	31.9	31.7	31.0
Plastic	4.1	4.4	4.9
Rubber, Leather, and Textiles	4.5	5.1	5.3
Wood	3.9	3.7	3.7
Organics (Food and yard waste)	35.5	35.1	35.0
Misc./Inorganics	1.6	1.8	2.3
	100.0	100.0	100.0

[a]Source: Franklin, W. et al., "Municipal Waste Generation and Composition to 1990," *Solid Waste Management and the Paper Industry,* Solid Waste Council of the Paper Industry, February 1979.

main cause of the variation in refuse quantities, as mentioned previously. The composition of the commercial and light industrial portion of MSW tends to be stable throughout the year.

ENERGY CONTENT

The generally accepted energy content of MSW is 4,500 Btu per pound on an as-received basis. This value is based on moisture, ash, and inorganic noncombustible materials contents of 25, 6, and 18.5 percent, respectively. As with the other waste characteristics, energy content is community specific and corresponds to variations in composition. Future changes in waste composition could affect the unit energy content (Btu per pound) of solid waste. Given the composition changes forecast in Table 3-2, potential energy losses in some material categories, such as paper, would be offset by gains in other categories such as plastics. The net effect is that the average unit energy content of MSW is anticipated to remain stable or to increase slightly through 1990.

The quantity of solid waste discarded in a community will determine the total amount of energy available for recovery. Due to seasonal composition changes, the unit energy content of solid waste also will vary by the month and season of the year. Given the nature of this variation (more yard waste and beverage containers in summer, less in winter), the average unit energy value of solid waste is lower in summer than in winter, all other factors being constant. Even so, since this variation is limited to residential solid waste which comprises only 67 percent of MSW, the overall variation in the unit energy content is minimal. Perhaps a more important seasonal variable is weather. Periods of high precipitation will casue the moisture content of MSW to rise, thus lowering the energy content on an as-received basis. Conversely, dry weather will cause the moisture content to be less than average, which would raise the as-received energy content of MSW.

4
MARKETS

INTRODUCTION

Securing a market for the recovered energy and material products is perhaps the single most important element in determining whether a resource recovery project will succeed or fail. This is because resource recovery is generally not economically competitive with conventional forms of waste disposal unless substantial cost offsets are derived from the sale of recovered products. Establishing economic viability through market revenues is so important that it is often said that markets drive the selection of technologies. While this may hold true for large-scale systems, it is less accurate for small-scale facilities, especially with regard to recovered material products. In such systems, the volume of wastes is insufficient to make separation of materials worthwhile. Nevertheless, special circumstances, e.g., a large local consumer of scrap metal, may warrant materials recovery.

In addition to understanding recoverable products and potential customers, it is important to recognize that the products must be aggressively marketed beginning in the project's earliest stages. The importance of this element is often overlooked since public agencies are unaccustomed to salesmanship. Potential markets not only must be identified but also must be firmly committed to the project early in the development process. This chapter presents a program for systematically identifying and committing potential markets. Because energy products—steam, electricity, and refuse-derived fuels—are most lucrative, primary emphasis is placed on these products, leaving material products with a more cursory treatment.

POTENTIAL MARKETS

A wide range of products can be recovered from solid waste. However, this range is somewhat limited when dealing with small- and medium-scale systems. For these systems, the most common products and their respective markets are examined—energy products first, then material products. Figure 4–1 graphically illustrates typical resource recovery system products.

SECONDARY PRODUCTS

PRIMARY PRODUCTS

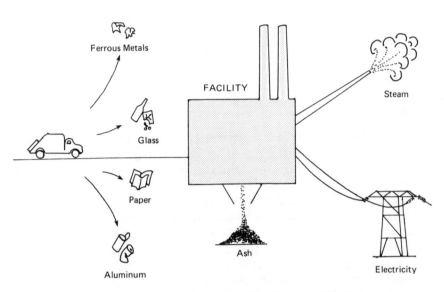

Figure 4–1. Typical resource recovery system.

Thermal Energy

Thermal energy is transportable through the use of a thermal transfer fluid. Waste-to-energy plants have utilized steam, hot water, chilled water, and hot air as thermal transfer fluids. The choice of thermal fluid depends upon the requirements of the system, particularly the end user of the energy. A recent survey of waste-to-energy plants indicated that steam was the preferred transfer fluid by a wide margin.[1] Of the 90 facilities for which definite markets for the energy were identified, 65 delivered steam to some end use. The end uses were: (1) space heat, 31 facilities; (2) process heat, 35 facilities; (3) mechanical energy, 1 facility; and (4) electricity generation, 36 facilities. Twenty-four facilities utilized the steam produced for two or more end uses; for this reason, there are more end uses than plants.

The choice of thermal transfer fluid is dependent upon the end use application of the 90 plants in the survey. Three are using hot water for space heating, six are using chilled water for space cooling, and one is using hot air for drying.

---

1. "Resource Recovery Activities," *City Currents,* U.S. Conference of Mayors, April 1984.

Steam

As the survey figures show, steam is the most prevalent thermal transfer fluid used in waste-to-energy facilities. This is in large part due to steam's efficiency and flexibility. The characteristics of the steam vary widely depending upon the end use. Steam temperature requirements range from 250 to 1,050°F, with pressures ranging from 150 to 500 psi. For electric power generation, highest efficiency is achieved by using steam at maximum temperature and pressure. The concept of generating both steam and electricity, or cogeneration, is discussed in the next section on electricity products and markets. In systems using solid waste as the sole or primary fuel, steam is seldom produced at more than 600 psi and 750°F (lower for small systems) in order to prevent slagging and corrosion of the boiler tubes. If necessary, the steam can be raised to the desired temperature and pressure in separate units. Table 4-1 shows the typical steam characteristics for the end uses discussed. The table has two entries for electrical generation. Low-pressure turbines are used when steam is produced from solid waste, due to the temperature and pressure restrictions mentioned. These result in some loss of efficiency when compared to the use of high-pressure turbines. An auxiliary boiler fired by fossil fuels can be used to raise the temperature and pressure of steam to that required by high-pressure turbines. This approach

Table 4–1. Typical Steam Characteristics.

	PSIG	°F
Space Heat		
Low Pressure	0–15	212–250
High Pressure	15	250
Process Application		
Washing	10–30	239–274
Hospital Equipment	40–50	287–298
Candy Manufacture	70–75	316–320
Laundry	100	338
Plastic Molding	125	353
Paper Manufacture	175	378
Electrical Generation		
Low-Pressure Turbines[a,b]	665	500
High-Pressure Turbines	1,800	1,000
Mechanical Energy[a]	15	250

[a]Turbine selection is a function of use and steam characteristics. Many choices are available utilizing saturated steam or supersaturated steam with 100 to 400°F of superheat.
[b]Due to corrosion in the boiler, refuse-fired systems utilize low-pressure turbines with a maximum temperature of 750°F.

has yet to be taken in waste-to-energy facilities but has been proposed by several small-scale vendors.

As mentioned, space heating is the main end use for steam produced in waste-to-energy plants. The 31 plants listed in the survey range in size from 31,500 to 1.35 million pounds of steam per day for the small-scale systems. Seven of these plants provide only steam for space heating. The remainder have at least one other end use. Steam for district heating is provided by five of the plants. The remainder provide steam to a single or primary end user. These include industrial firms, military complexes, colleges and universities, governmental office complexes, and hospitals.

The specifications for steam recovered from solid waste depend entirely on the needs of the market. The following criteria are important in assessing steam markets:

- *Proximity to customers.* A steam-generating facility must have a nearby market because steam cannot be economically transported more than 1 to 2 miles. In congested areas, expensive pipeline installation may further restrict this distance.
- *Value.* The cost at which the steam is delivered must be competitive with the cost of the customer's alternative energy sources.
- *Quantity.* Price is affected by the ability of the steam plant to supply amounts of steam which are compatible with the customer's needs. If supply is guaranteed to the customer and peak loadings cannot be met entirely by burning refuse alone, then standby, fossil fuel–fired boilers will be needed and the price correspondingly increased.
- *Operating schedule.* The steam-producing facility must set up an operating schedule that conforms to that of the steam customer. This affects the price of the steam as well as the existence of the market itself.
- *Steam quality.* The temperature, pressure, and saturation at which the steam is produced must be within the limits acceptable to the customer, according to the steam contract. Variations from this norm could seriously affect the price of the steam.
- *Reliability.* If service is to be noninterruptible, contingency plans should be made when the resource recovery unit is out of service.
- *Timing.* This aspect can seriously affect the steam plant and the expected revenues. Unanticipated delays in construction of the facility could force the steam customer to secure another source.
- *Diurnal and seasonal fluctuations in demand.* Refuse-fired steam generators are best suited to generating a constant level of steam; if variations in demand do exist, they should be relatively predictable.

Steam can be marketed in two ways: as a guaranteed supply (uninterruptible service) or as a limited supply that requires a backup system (interruptible service). The pricing structure will vary in accordance with the type of service offered. In the first case, the resource recovery system provides a complete and reliable supply of steam and assumes the responsibility of producing steam from other sources if there should be an interruption in the production of steam from solid waste. If the resource recovery system is supplying steam that the customer does not have the capabilities of producing, then it must guarantee reliability of supply. When the resource recovery system's costs go up to meet the customer's criteria, so does the value of the steam it is selling. This steam has a value equivalent to what the customer would have to spend to produce it. In the second case, the customer buys all the steam produced from solid waste and assumes the burden of producing additional steam in the event that this supply is interrupted or is not adequate to meet its demand. In this case, the value of the steam is necessarily lower in that it is limited to the value of the fuel saved by the customer. At the same time, the development agency assumes less risk and responsibility.

A major constraint in selling steam based on municipal solid waste as feedstock is that the generation of steam in a municipal refuse incinerator is not easily varied to meet load conditions. In cases where the steam-generating value of the solid waste exceeds the demand, the plant must be equipped with a boiler bypass flue or steam-condensing equipment. Conversely, when the demand for steam exceeds the supply, the plant must be equipped with an auxiliary fuel-fired system if its customer service contract(s) demands an uninterrupted supply of steam. Experience has shown that the additional cost of this equipment is only justified where firm markets can be secured for all the steam produced. To date such markets have not materialized to any extent.

Alternatively, a small refuse-fired unit could be constructed as part of a larger steam-generating system and feed into the system what steam it had to sell. When this unit did not receive sufficient refuse, or when the heating value of the refuse was low due to adverse weather conditions or composition, it simply would not produce steam.

The potential end use customers for steam fall into two general categories: (1) new requirements and (2) conversion of existing boilers or requirements to resource recovery plants. The need for energy of a new industrial, institutional, etc., facility is straightforward. Replacement of existing boilers can be achieved for the following reasons: (1) lower fuel cost over life of facility, (2) stable fuel supply, (3) air pollution requirements, or (4) normal life cycle replacement. Therefore, existing boilers constitute a market for resource recovery plants.

Precise data on the total number of boilers and their related capacities are unavailable. However, a recent (1979) estimate was developed for the U.S.

Environmental Protection Agency (EPA) by PEDCo to estimate the emissions from industrial, commercial, and industrial boilers. According to the data in the PEDCo report, there were about 1.8 million industrial, commercial, and institutional boilers in operation in 1978. Of these boilers, almost 1.3 million were small units used primarily for space heating in commercial and institutional buildings. The remaining 0.5 million units were industrial boilers used for process steam and space heating. These industrial boilers had a total nameplate capacity of 3.1 trillion Btu per hour.

In 1978, there were an estimated 28,925 industrial, commercial, and institutional boilers in operation that had a nameplate capacity of more than 25 million Btu per hour. Table 4–2 presents a breakdown of the number of boilers and nameplate capacity for units that range in size from more than 25 to 50, more than 50 to 100, and more than 100 million Btu per hour.

Small-scale waste-to-energy systems could produce steam at the same rate as conventional fuel-fired boilers with nameplate ratings of 25 to 125 million Btu per hour, based on an operating rate of 80 percent of capacity. Boilers in the more than 100 to 125 million Btu per hour range could not be separated from the data as presented in the PEDCo report. Therefore, the data in the first two rows of Table 4–2 understate the number and capacity of boilers which could be affected by small-scale energy recovery systems.

The actual ability of waste-to-energy systems to replace boilers at various rated capacities will be a function of specific load management demands. For the most part, however, small-scale energy recovery facilities provide the opportunity for significant replacement of conventional fuel boilers with ratings up to 125 million Btu per hour.

The total amount of usable energy that could have been recovered (50 percent conversion efficiency) from solid waste in 1981 was 75,909 million Btu per hour. By comparison, the total nameplate capacity of industrial, commercial, and institutional boilers rated at greater than 25 million Btu per hour was 2,502,700 million Btu per hour in 1978. The maximum energy out-

Table 4–2. Distribution of Industrial, Commercial, and Institutional Boilers by Size.[a]

SIZE (10^6 BTU/HR)	NUMBER OF BOILERS	TOTAL NAMEPLATE CAPACITY (10^9 BTU/HR)
25 to 50	16,483	608.7
50 to 100	6,840	503.0
> 100	5,602	1,391.0
TOTAL	28,925	2,502.7

[a]Source: PEDCo, "Population and Characteristics of Industrial/Commercial Boilers in the United States," U.S. Environmental Protection Agency, Research Triangle Park, N.C., 1979.

put of solid waste represents almost 4 percent of productive capacity of the target boiler market (more than 25 million Btu per hour rating), when operated at 80 percent of the nameplate rating. This indicates that on a nationwide basis, the capacity for steam production greatly exceeds the amount that could be produced using solid waste.

Actual steam produced by boilers is less than the nameplate capacity due to redundancy, shutdown during plantwide vacations, seasonal fluctuations, excess capacity, and outmoded equipment. A 1980 estimate of steam demand developed for the U.S. Department of Energy (DOE) set annual steam demand at 3,934,000 MBtu.[2] On an hourly basis, this converts to 449,100,000 Btu, resulting in a utilization factor of 18 percent. Converted to steam, the 75,900,000 Btu of usable energy available from solid waste could have supplied 16.9 percent of the estimated 1981 steam demand.

SIC Code Distribution.   Steam demand for the manufacturing Standard Industrial Codes (SIC) amounted to 3,694,000 million Btu in 1980, according to DOE estimates, Table 4–3. Forecast steam demands for 1985 and 1990 for the same SIC codes are 4,911,900 million Btu and 5,918,600 million Btu, respectively. Industrial manufacturing steam demand greatly exceeds the available supply of energy from municipal solid waste (MSW). The data in Table 4–3 understate actual steam demand, since the nonmanufacturing industrial sectors (agriculture, mining, and construction) and commercial/institutional steam outputs were excluded.

Table 4–3. Industrial Manufacturing Steam Demand by SIC Code, 1980–1990.[a]

SIC CODE	INDUSTRY	ANNUAL STEAM DEMAND (10^{12} BTU)		
		1980	1985	1990
20	Food	374.1	446.0	510.6
21	Tobacco	1.6	1.7	1.9
22	Textiles	108.8	139.1	173.6
23	Apparel	7.7	9.7	10.9
24	Lumber, Wood	4.1	5.4	5.8
25	Furniture	3.8	5.0	5.6
26	Paper	1,194.4	1,587.8	1,846.9
27	Printing	2.6	3.3	3.8
				(cont.)

2. Energy and Environmental Analysis, Inc., "Market Oriented Progress Planning Study—Industrial Sector," U.S. Department of Energy, Washington, D.C., December 1977.

Table 4–3. (continued)

SIC CODE	INDUSTRY	ANNUAL STEAM DEMAND (10^{12} BTU)		
		1980	1985	1990
28	Chemicals	1,270.5	1,818.8	2,396.4
29	Petroleum Refining	498.9	589.6	605.7
30	Rubber and Miscellaneous Plastics	57.7	84.2	106.3
31	Leather	3.9	4.3	4.7
32	Stone, Clay, and Glass Products	0.0	0.0	0.0
33	Primary Metals	74.8	97.3	105.0
34	Fabricated Metal Products	17.6	22.2	25.0
35	Machinery	19.9	26.3	32.0
36	Electrical Machinery	13.2	18.1	21.8
37	Transportation Equipment	32.2	42.2	49.4
38	Measuring Equipment	4.8	6.4	7.8
39	Miscellaneous Manufacturing	3.4	4.5	5.4
	Total Manufacturing	3,694.0	4,911.9	5,918.6

[a]Source: Energy and Environmental Analysis, Inc., "Market Oriented Progress Planning Study—Industrial Sector," U.S. Department of Energy, Washington, D.C., December 1977.

Industries in which steam demand is high (greater than 50 trillion Btu per year) are:

- Food
- Textiles
- Paper
- Chemicals
- Petroleum refining
- Rubber and miscellaneous plastics
- Primary metals

The food industry generally is an undesirable market for energy from solid waste because of the seasonal demand for steam. During the fall harvest, demand is high, followed by a declining demand as food stock is processed. Another unlikely energy market is the paper industry. Pulp and paper mills typically are located near the raw material resources—trees. Most forests are located away from the population concentrations necessary to justify a waste-to-energy system using solid waste. However, an energy recovery system that would use a combination of municipal solid waste and industrial waste is feasible. A system of this type is in operation in Groveton, New Hampshire, 24 TPD (tons per day).

The City of Springfield, Massachusetts is currently implementing a distributing heating system utilizing high-temperature hot water. This is currently in the procurement phase and is partially funded by a U.S. Department of Housing and Urban Development (HUD), Urban Development Action Grant (UDAG). This system, when complete, will obtain part of its energy from a waste-to-energy plant.

Hot Water

In the waste-to-energy survey, three plants produced hot water as a thermal transfer fluid. These supply space heating end users. The Disney World facility, the Akron, and the Toronto, Ontario plants produce hot water. The temperature of the water in these systems ranges from 200 to 320°F, with pressures in the range of 60 to 250 psig.

High-temperature hot water is an innovation in the United States; however, it has been used in Europe for district heating since the Second World War. The facilities in Europe in 1983 provided 3.7 million megawatts of installed district heating capacity.[3] High-temperature hot water has three distinct advantages over steam: (1) greater transport distances, (2) cheaper piping costs, and (3) more accurate metering. High-temperature hot water can be economically transported long distances in the current state-of-the-art piping. Steam has a limitation of approximately 2 miles. The constraint for the transport distance of hot water is primarily economic. Feasibility is determined by an analysis which looks at cost of energy to the system and competing sources. This is then traded off against cost of piping and distance of transport. Certain projects in Canada and Sweden, where the cost of heat to the system is very low, have had favorable economics with 40 mile transport systems. This is unusual; a nominal maximum is about 15 miles. Piping systems utilizing low carbon steel and polyurethane insulation are lower in cost than comparable steam piping. Also, the need for condensation traps is eliminated, which results in additional savings. Further, a very accurate measurement of the amount of heat used by each customer is easily obtained with an input and output temperature sensor and a flowmeter coupled to a microprocessor. These reliable devices can be interrogated remotely, thereby eliminating the need for manual reading.

Industrial use of hot water is currently limited primarily to the food industry. Also, hot water is used for space heating by industrial, commercial, and institutional facilities utilizing fossil fuels. Although no estimate of the number of these small-scale facilities is available, they represent a large potential market for energy recovered from waste. This market could be supplied either by dedicated boilers or by district heating systems.

---

3. *District Heating Handbook,* International District Heating Association, Washington, D.C., 1983.

Hot Air

The survey shows only one facility producing hot air as a thermal transfer medium: Harrisburg, Pennsylvania utilizes hot air to dry its sewage sludge. Sludge drying could utilize large amounts of energy from waste.

Hot air is used by industry to dry products such as lumber and bricks, as well as to melt materials such as glass. Typical processes rely upon the heat released from natural gas burners. A heat exchanger can also be used, particularly in drying applications.

In the case of hot air, manufacturing industrial demand was 1,711.0 trillion Btu in 1980, Table 4-4. Stone, clay, and glass products and primary metals account for 67 percent of this demand. The table includes all types of hot air uses, some of which probably would be inappropriate for solid waste-fired applications due to the need for a clean hot air stream. Even so, conversion of all the solid waste discarded in 1981 to hot air would have produced 664.9 trillion Btu in a small-scale waste-to-energy system using a heat exchanger, based on a conversion efficiency of 50 percent. Hot air demand is forecast to rise to 2,233.2 and 2,529.0 trillion Btu in 1985 and 1990, respectively.[4] Solid waste would be able to supply 31 and 29 percent of demand in these years by the conversion of the total supply to hot air.

Table 4-4. Industrial Manufacturing Hot Air Demand by SIC Code, 1980-1990.[a]

		ANNUAL HOT AIR DEMAND (10^{12} BTU)		
SIC CODE	INDUSTRY	1980	1985	1990
20	Food	75.9	90.6	103.6
21	Tobacco	5.0	5.4	5.9
22	Textiles	23.0	29.4	36.7
23	Apparel	17.5	21.9	24.7
24	Lumber, Wood	26.8	35.3	38.3
25	Furniture	9.3	12.2	13.5
26	Paper	82.1	109.1	126.9
27	Printing	10.9	13.6	15.7
28	Chemicals	52.0	74.4	98.0
29	Petroleum Refining	0.0	0.0	0.0
30	Rubber and Misc. Plastics	18.1	26.3	33.2
31	Leather	3.4	3.7	4.1

4. Energy and Environmental Analysis, Inc., "Market Oriented Program Planning Study—Industrial Sector," U.S. Department of Energy, Washington, D.C., December 1977.

Table 4–4. (continued)

SIC CODE	INDUSTRY	ANNUAL HOT AIR DEMAND (10^{12} BTU)		
		1980	1985	1990
32	Stone, Clay, and Glass Products	451.1	579.6	662.8
33	Primary Metals	700.4	921.8	1,000.7
34	Fabricated Metal Products	73.3	92.5	104.2
35	Machinery	60.8	80.7	98.1
36	Electrical Machinery	36.9	50.5	61.0
37	Transportation Equipment	52.7	69.1	80.9
38	Measuring Equipment	5.9	8.0	9.8
39	Misc. Manufacturing	6.9	9.1	10.9
	Total Manufacturing	1,711.0	2,233.2	2,529.0

[a]Source: Energy and Environmental Analysis, Inc., "Market Oriented Program Planning Study—Industrial Sector," U.S. Department of Energy, Washington, D.C., December 1977.

Supply/Demand Considerations

At large, the market for thermal energy varies with the season of the year. This is clearly the case for space heating. Also, a number of industrial processes are seasonal such as food processing. In order to maximize revenues and to provide for disposal of the solid waste which is generated every day throughout the year, a thermal plant should attempt to balance the thermal load by selecting its end uses. For example, by providing chilled water as well as steam, a waste-to-energy plant can level demand. Also, an industrial process which can accept any surplus thermal energy displacing fossil fuel is ideal. Of course, the price per Btu of the latter user is going to be low. Good management practice will schedule routine maintenance and capital replacement during the seasonal period of lowest demand.

Many thermal energy end use customers are likely to rely only on the waste-to-energy plant for their supply. Redundant capacity will have to be included in the waste-to-energy project. This may include fossil fuel backup. These considerations affect the contractual arrangements between the plant and the end user, in addition to the higher price for the energy. Contracts must cover the event of unscheduled outage causing an interruption of service.

Electricity

The *City Currents* survey of waste-to-energy plants indicates that 36 out of the 90 surveyed are producing electricity. The local utility provides the market for the electricity from 31 of these plants. Four plant operators utilize the electricity within their complexes; these include universities, military bases, and municipalities. The refuse fuel plant in Haverhill/Lawrence, Massachusetts provides electricity for industrial use, allowed by a special state law.

Specifications for the sale of electricity are dictated by the customer. If the customer is a utility (as most are in the survey), the characteristics—voltage and number of lines—of the nearest transmission line will dictate specifications. Also, the utility will most likely either specify or require approval of the interconnection design.

Due to the economies achievable through large-scale generation of electricity by utilities, it is difficult for small waste-to-energy facilities to compete in the production of base load power. Although such production is technologically indistinguishable from conventional methods, the economics remain a major barrier to implementation.

In many states, electricity recovered from a resource recovery facility may be sold only to the electric utility serving the area, because within that service area the utility can be exempted from competition. The sale of electricity to another utility could be regulated by the Federal Energy Regulatory Commission (FERC) if the utility's grid extends beyond the state's borders. The price that a utility will pay for electricity depends on whether it is used to satisfy base load or peak load demand. Peak load marketing commands a much higher price (perhaps three to five times the price of base load electricity); however, a resource recovery facility needs to sell base load electricity in order to maintain a continuous solid waste disposal operation.

A project sponsor considering the sale of electricity to other utilities may seek to establish a floating price for electricity, whereby the price per kilowatt-hour rises as the demand on the utility increases. Thus, the price would be a function of the incremental direct costs the utility incurs in producing the electricity needed to meet increased demand. Another approach would be for the resource recovery facility to sell the electricity to the utility at a price equal to its average cost of production.

It is desirable to transmit electricity from a resource recovery facility directly to the market. Thus, the facility should be located near the market or an access point on the market's grid. If the market is remote from the facility, then the power can be "wheeled" through the local utility's grid. However, a "wheeling" arrangement may incur a "wheeling" charge.

Many of these pricing considerations become irrelevant in light of recent federal legislation which, in effect, establishes floor prices for electricity generated by certain kinds of facilities. The implications of this legislation will be discussed.

In 1980 there were two waste-to-energy plants with electrical generation capacity. Now 35 are either operating or in the process of being implemented. One of the reasons for this lies in the incentives provided by Sections 201 and 210 of the Public Utility Regulatory Policies Act of 1978 (PURPA). Additional financial incentives are also contained in the Energy Tax Act of 1978, the Crude Oil Windfall Profits Tax Act of 1980, and for natural gas-fired cogenerators, the Natural Gas Policy Act of 1978. Further, the Power Plant and Industrial Fuel Use Act contains provisions whereby major fuel burning installations may receive exemptions from prohibitions on burning natural gas or oil if they are cogenerators.

Section 210 of PURPA requires the Federal Energy Regulatory Commission to issue rules for the encouragement of cogeneration and small power production pursuant to the Act. These rules include assurance that qualifying cogenerators and small power producers receive prices for sales to electric utilities which are just and reasonable to the ratepayers of the electric utility, nondiscriminatory toward the cogenerator and small power producer, and in the public interest. These prices are termed "avoided costs" and are the incremental costs the utility would have experienced if it had generated an equivalent amount of electric energy and capacity itself or had purchased it elsewhere.

Under FERC rules, a cogeneration facility is a facility that produces both (1) electric energy and (2) steam or forms of useful energy (such as heat) that are used for industrial or commercial, heating or cooling purposes. FERC has issued efficiency standards for cogeneration facilities to qualify under this program. Aside from these efficiency standards, there is no size limitation. However, cogenerators do not qualify if utilities own more than 50 percent of the equity interest in the facility.

A small power production facility is a facility that produces no greater than 80 megawatts of electric energy solely by the use, as a primary energy source, of biomass, waste, renewable resources, or any combination thereof. Wood and wood products are classified as biomass so that, for example, a generating facility using by-products of the wood, paper, and pulp industries may qualify if (1) it is not greater than 80 megawatts; (2) it does not use oil, gas, or coal for more than 25 percent of its fuel; and (3) not more than 50 percent of the equity interest is owned by electric utilities.

The FERC regulations provide a number of advantages for qualifying cogeneration and small power production facilities. First, the rules under Section 210 provide for exemption from federal and state public utility regulation of those facilities that qualify under the Commission's rules. Therefore, a qualifying cogenerator or small power producer is generally exempt from the federal and state laws that regulate the financial or organizational arrangements of public utilities and the rates at which they can sell electric energy, as well as from the Public Utility Holding Company Act.

Second, qualifying cogenerators and small power producers are assured of a market for their power—a market that may not have existed prior to the implementation of Section 210 of PURPA. FERC rules require that each electric utility purchase electricity from these facilities.

Third, the costs incurred by the cogenerators or small power producers to generate electricity are irrelevant to what the utility will have to pay for it. The sole determinant for the price for the power will be what the utility would pay based upon the avoidance of generating costs on its system by reason of the purchase of power from the cogenerator or small power producer.

Section 210 of PURPA called for state implementation of regulations within one year of the effective date of issue of FERC's regulations, which occurred March 20, 1981.[5] Each state regulatory authority had to determine the nature and extent of the electric utility system cost data upon which avoided costs to utilities will be calculated. These data had to be filed by utilities with state authorities by November 1, 1980. All states have complied.

Paragraphs (b)(1)–(3) of Section 210 offer FERC suggestions concerning the nature of the data to be submitted. However, a state authority could adopt an alternative as long as it determines that avoided costs can be calculated from those data. Any state deciding to accept alternate data must notify the FERC within 30 days of making such a determination. In that these data will form the basis of the calculation of avoided costs and, ultimately, the price that a cogenerator or small power producer will receive for power, it behooves cogenerators and small power producers to participate in the rate-setting process. Finally, each state was required to implement PURPA regulations. It should be stressed that the FERC gave the states great flexibility in the manner of implementation they must follow, leaving latitude to the states to fit their implementation into the context of their particular organizational structure.

Industrial Energy Market Summary

Energy consumption by manufacturing industries is a substantial and untapped market. In 1978, for example, this segment of the economy consumed 8,928.2 trillion Btu of energy in the form of fuel oil, natural gas, and coal, as well as 2,304.5 trillion Btu of electricity, Table 4–5. The three fossil fuels (oil, natural gas, and coal) are of primary importance as energy resources for which MSW could substitute. The data in Table 4–5 show that for the fossil fuels, consumption ranged from 145.2 trillion Btu in Region VIII to 2,347.2

---

5. For a status report on state implementation of PURPA, see "Survey: PURPA 210 Implementation," by Ann Cavan, *Hydro-Review,* Winter 1984.

Table 4-5. Manufacturing Industrial Energy Consumption by Region and Type, 1978.[a-c]

| | FUEL OIL | | | | NATURAL GAS | | COAL | | PURCHASED ELECTRICITY | | |
| | RESIDUAL | | DISTILLATE | | | | | | | | |
REGION[d]	QUANTITY (10^12 BTU)	PERCENT	QUANTITY (10^12 BTU)	PERCENT	QUANTITY (10^12 BTU)	PERCENT	QUANTITY (10^12 BTU)	PERCENT	QUANTITY (10^12 BTU)	PERCENT	TOTAL (10^12 BTU)
I	178.5	54.5	34.9	10.7	44.2[i]	13.5	0.07[j]	0.0	69.8	21.3	327.4
II	211.8	32.2	61.7	9.4	162.8	24.7	63.5	9.6	158.7	24.1	658.5
III	243.6[e]	19.7	94.7[e]	7.6	397.8	32.1	273.2[k]	22.0	230.1	18.6	1,239.4
IV	376.3	19.8	109.0	5.7	610.6	32.1	286.9[l]	15.1	517.6	27.2	1,900.4
V	260.6	9.6	106.0	3.9	1,305.8	47.8	487.2	17.8	570.1	20.9	2,729.7
VI	148.6[f]	5.6	31.4[f]	1.2	2,130.8	80.0	36.4[m]	1.4	315.3	11.8	2,662.5
VII	24.9	5.0	26.8	5.3	273.1	54.3	89.1	17.7	89.1	17.7	503.0
VIII	11.4[g]	6.2	8.2	4.5	119.8	65.1	5.8[n]	3.2	38.9	21.1	184.1
IX	52.0[h]	8.2	19.8[h]	3.1	395.6	62.3	23.1[o]	3.6	144.0	22.7	634.5
X	44.0	11.2	16.8	4.3	161.5	41.1	0.0[p]	0.0	170.9	43.5	393.2
TOTAL	1,551.7		509.3		5,602.0		1,265.2		2,304.5		11,232.7

[a] Compiled from data in "1978 Annual Survey of Manufacturers—Fuels and Electricity Consumed," U.S. Bureau of the Census, Washington, D.C., August 1981.
[b] Includes data for only the 50 states and the District of Columbia.
[c] Includes those industrial activities defined by SIC codes 20 to 39 (manufacturing) only.
[d] Federal government regions.
[e] Excludes Maryland.
[f] Excludes Oklahoma.
[g] Excludes Montana.
[h] Excludes Nevada.
[i] Excludes Vermont.
[j] Excludes Connecticut, Maine, Massachusetts, and New Hampshire.
[k] Excludes Delaware.
[l] Excludes Mississippi.
[m] Excludes Louisiana, New Mexico, and Oklahoma.
[n] Excludes Colorado, Montana, North Dakota, South Dakota, and Wyoming.
[o] Excludes Arizona.
[p] Excludes Idaho, Oregon, and Washington.

trillion Btu in Region VI. Purchased electricity consumption by manufacturing industries in 1978 ranged from 38.9 trillion Btu in Region VIII to 570.1 trillion Btu in Region V. Fuel consumption in each region is generally a function of geographic size, industrialization, and climatic conditions.

There also was significant variation in the type of fuel consumed by region. Residual oil, for example, accounted for 69.3 percent of the energy mix in Region I and 6.3 percent in Region VI. The energy mix in an area with a sufficient quantity of waste to justify recovery is critical. Recovery revenue depends on the type and cost of fuel displaced. Fuel oil, which currently is the highest cost fuel per million Btu, is the preferred energy resource for substitution. Natural gas also will be a viable fuel for substitution as price controls are removed through 1985. Coal is a low-cost fuel, which makes substitution by MSW non–cost effective in most cases.

The trend in industrial energy consumption is towards increased use of coal and electricity, while use of fuel oil is forecast to decline (1980 Annual Report to Congress, U.S. Department of Energy, Washington, D.C., 1981). Coal consumption accounted for 23.5 percent of industrial energy in 1978 and is forecast to grow to 41.2 percent by 1990. During this same period, fuel oil use is forecast to decline from 17.9 to 3.0 percent. These data developed by the U.S. Department of Energy cover four industrial sectors: (1) agriculture, (2) mining, (3) construction, and (4) manufacturing.

Cogeneration

A cogeneration facility is one which produces both steam and electricity as salable products. As mentioned earlier, the survey lists 35 facilities that produce electricity. Of these, 15 also produce steam as a product and are, therefore, cogeneration plants. The advantage of cogeneration is its increased efficiency of fuel utilization. When electricity alone is produced, 25 to 35 percent of the energy in the fuel is captured as product. In a cogeneration plant, this is increased to 65 to 84 percent depending upon plant design, fuel, and other factors.

This produces two benefits in a waste-to-energy facility. These are:

1. increased revenue from product sales and
2. potential to optimize energy production levels.

The realization of this latter benefit is dependent upon the needs of the energy customer as reflected in the market agreements.

REFUSE-DERIVED FUEL
AND MATERIALS RECOVERY

Refuse-Derived Fuel

The survey of waste-to-energy plants indicates that ten plants are refuse-derived fuel (RDF) producers. The characteristics of RDF vary considerably as indicated by ASTM definitions shown in Table 4-6. Several of the plants listed in the survey produce RDF with proprietary names and specifications. These include 300 Fuel™, Processed Refuse Fuel (PRF), Organi-FUEL 100, and K-Fuel. In addition, several facilities producing thermal products or electricity for end use customers burn prepared solid waste or RDF in these furnaces.

A number of large-scale RDF production plants have been forced to close. This has generally involved problems concerning the fuel specifications. These have involved feeding RDF to existing boilers as a supplemental fuel along with the original specification fuel for the boiler. The Milwaukee, Wisconsin facility is an example.

The processing of solid waste into RDF generally is not considered applicable to small throughput tonnages. However, since several small-scale

Table 4-6. Refuse-Derived Fuel Definitions.[a]

Refuse-derived fuel (RDF) is defined as fuel extracted from solid waste to be used for combustion processes or as feedstock to other process systems. The forms of fuel derived from municipal solid waste include:

RDF-1 Wastes used as a fuel in as-discarded form with only bulky wastes removed.

RDF-2 Wastes processed to coarse particle size with or without ferrous metal separation.[b]

RDF-3 Combustible waste fraction processed to particle sizes—95 percent passing 2 inch square screening.[c]

RDF-4 Combustible waste fraction processed into powder form—95 percent passing 10 inch mesh screening.

RDF-5 Combustible waste fraction densified (compressed) into the form of pellets, slugs, cubettes, or briquettes.

RDF-6 Combustible waste fraction processed into liquid fuel.

RDF-7 Combustible waste fraction processed into gaseous fuel

[a]Source: *Thesaurus on Resource Recovery Terminology* (Herbert I. Hollander, editor), ASTM Special Technical Publication 832, Philadelphia, 1984.
[b]Shredded refuse fuel, used principally as a supplement in utility or industrial boilers which have ash handling capabilities. By means of a separation system, much of the metal, glass, and other inorganics is first removed. RDF is the remaining organic fraction which has been processed to relatively uniform size particles.
[c]A shredded fuel derived from municipal solid waste (MSW) that has been processed so as to remove metal, glass, and other entrained inorganics. The material has a particle size such that 95 weight percent passes through a 2 inch (50 mm) square mesh screen. RDF-3 is used as a primary or supplementary fuel in existing or new, industrial or utility boilers.

RDF systems are currently operating in the United States (Ames, Iowa—150 TPD; Madison, Wisconsin—250 TPD), the products and markets merit brief examination.

RDF can be used as the solid fuel in dedicated boilers or as a supplemental fuel in large spreader-stoker or suspension-fired boilers. RDF is usually sold on the basis of its energy content (Btu per pound), and its price is indexed to the price of a competing energy source, usually coal. Additional costs in using the fuel (storage facilities, feeding equipment, additional maintenance, etc.) usually are discounted from the price of the RDF. Under ideal conditions, RDF is manufactured in a facility adjacent to, or integrated with, the market so that it can be conveyed or piped pneumatically to the storage bins/boilers. Where this is not possible, transport is normally via truck or rail.

The processing of solid waste into RDF is usually accompanied by the removal of noncombustible materials for subsequent resale. In the following section, we address the characteristics of these products and their markets.

Ferrous Metals

The survey of 99 resource recovery plants indicates that 31 either recover or plan to recover ferrous metals. These include both RDF production plants and mass burning plants where the ferrous metals are recovered from the ash. The recovered ferrous metals in the latter case generally have a lower economic value.

It is worth noting that the amount of ferrous metals in solid waste may be declining. This is due primarily to changes in the packaging industry. Aluminum has been replacing steel in the beverage can market, and there is currently market pressure from aluminum and other materials on food bimetal cans.

Ferrous metals, excluding automobiles, make up about 8 percent of solid waste. About 50 percent of these ferrous discards are in the form of cans. The remainder consists of appliances (16 percent) and miscellaneous items such as hardware, metal castings, and nondescript pieces of metal (33 percent). In 1979, only about 500,000 tons, or 4.4 percent, of the ferrous metal in solid waste was recycled.

The characteristics of steel cans bear heavily on the marketability of ferrous metals recovered from solid waste. Scrap steel cans can be marketed to three industries: steel, copper precipitation, and detinning. All recovered steel scrap potentially can be marketed to the steel industry, while only scrap steel cans are generally acceptable to copper precipitation and detinning markets.

The steel industry represents the largest potential market for ferrous metals recovered from solid waste. Eighty-nine companies operated 150 plants in 1979. Ninety-five percent of these plants are located in Standard Metropolitan Statistical Areas (SMSAs) and are, therefore, in close proximity to potential recovery sites.

Closely allied to the steel industry is the foundry industry which melts pig iron scrap for casting. There are 4,000 to 5,000 iron and steel foundries in the United States which consume about 28 percent of the nation's purchased scrap. To date there has been relatively little experimentation with high can content municipal ferrous scrap by this industry.

Despite the use of almost 40 tons of purchased scrap by the steel industry in 1980, the amount of municipal ferrous scrap consumed was insignificant. Instead, the scrap used consisted primarily of borings, stampings, and turnings from fabrication operations (e.g., automobile or can manufacturing), demolition steel, abandoned automobiles, and postconsumer scrap from a variety of industrial sources. The quantity of steel recovered from solid waste (excluding automobiles) constitutes only about 0.1 percent of the steel industry's present scrap consumption. Municipal scrap is significantly different from other sources of ferrous scrap. The lead and tin in the ferrous scrap are contaminants in steelmaking. The scrap may also contain organics or other materials that make it undesirable. Thus, it presently is an experimental input on an industrywide basis.

The steel industry's potential assimilative capacity exceeds the relatively small quantities of ferrous scrap from solid waste as indicated in the National Center for Resource Recovery (NCRR) study.[6] However, since the steel industry is in the early stages of experimenting with this type of scrap, the potential interest and demand are uncertain. The quality of municipal ferrous scrap will be key to attracting the steel industry to the solid waste market.

The copper precipitation industry, located close to copper mining operations in the Rocky Mountains, utilizes ferrous scrap as precipitation iron. In the process, the scrap is placed in a solution of copper sulfate and, in a chemical reaction, the copper is displaced by the iron, thus forming iron sulfate, while the copper is precipated and extracted. Despite the high transportation costs associated with the remote Rocky Mountain location, the industry regularly receives scrap from as far away as Chicago and St. Louis. However, the potential of this market is limited by the size of the industry and the current low demand for copper.

Detinners chemically process "tinplate" (such as that on cans) to remove tin content. Tin, which makes up roughly 0.4 percent of a "tin can," is a

6. E. Joseph Duckett, *Contaminants of Magnetic Metals Recovered from Municipal Solid Waste,* National Center for Resource Recovery, Washington, D.C., November 1977.

valuable commodity worth roughly $3–4 per pound. Although detinners process tinplate primarily for the tin content, the resulting "tin-free" steel is also valuable as a raw material to the steel or copper precipitation industries.

The ability to market ferrous metals from solid waste is heavily dependent on the form and purity of the recovered scrap. The three major potential markets have specification requirements that differ markedly. Therefore, to recover and sell municipal steel scrap successfully, it is essential to arrange a market prior to the selection of recovery configuration and technology.

Glass

The RDF production or processed waste plants in the survey indicate six that claim glass as a recovered product. Of these, only Wilmington, Delaware and Dade County, Florida are in current operation.

Glass constitutes about 10 percent of the municipal postconsumer waste stream in the United States and, in 1979, totaled 13.6 million tons. Glass containers represent the major portion of glass found in solid waste. Approximately two-thirds of these glass products are made of flint or clear glass. The remainder is split between amber glass used for beer bottles and green glass used for wine and soft drinks. In 1979 only 641,000 tons, or 5 percent, of the glass in postconsumer solid waste was recovered, overwhelmingly through recycling (source separation) systems.

There are two major potential markets for recovered waste glass: (a) as cullet for making new bottles and (b) as a raw material for making secondary products, e.g., glassphalt highway paving material, foamed insulation, or construction materials.

The most pressing market issue for recovered glass centers on the quality of the cullet. If the glass is properly sorted by color and if contaminants are kept to a minimum, it is likely that a buyer can be found. However, these are major barriers.

Color-mixed cullet has a more limited market potential than color-sorted glass. Color-mixed cullet is practically never used in making clear glass. Since roughly two-thirds of the industry's production is in clear glass, the potential buyers for mixed color cullet are limited. However, this limit has not been approached.

Glass manufacturers have established stringent contamination specifications for waste glass. Although pilot plant work has been performed, there are currently no full-scale glass recovery systems operating. Thus, it is uncertain whether these specifications can be met on a day-to-day production scale. The uncertainty regarding the ability to meet glass industry specifications should be investigated with potential buyers.

Aluminum

The survey of resource recovery plants indicates that ten facilities claim or plan to recover aluminum. However, it is not clear that the aluminum recovery portion of the operating facilities is working.

The primary form of aluminum recovery from solid waste currently practiced is source separation of aluminum cans through volunteer community recycling centers. In 1980 about 300,000 tons of aluminum cans were recovered, which represents 33 percent of all aluminum discarded. Source separated aluminum cans generally can be remelted by the primary producers and made directly into can stock.

Mechanical extraction of scrap aluminum from mixed solid waste is being developed and as yet has not been demonstrated on a commercial scale. It is anticipated that aluminum scrap extracted by mechanical means will be lower in quality than that recovered through source separation. It is likely that a large portion of this may be sold to secondary smelters who will pretreat and upgrade the aluminum by removing contaminants and dilute the alloy contents to acceptable levels. As is common with any scrap metal, the value of this recovered aluminum scrap is likely to be negotiated based on the quality of the recovered product and the specifications required by the purchaser.

Aluminum constitutes about 0.7 percent of the solid waste stream. About half of the aluminum discards in solid waste consists of cans, one-third is foil, and the remainder is largely from major appliances. Aluminum composition varies significantly from one community to another due to differences in aluminum beverage can distribution. Aluminum scrap constituted 24 percent of the aluminum produced in 1979. Sixty percent of the scrap utilized was consumed by secondary smelters, seventeen percent by primary producers, and the remainder by aluminum fabricators and foundries. There are 31 primary aluminum producers and 111 secondary aluminum smelters in the United States.

Nonferrous Metals

"Nonferrous metals" is an inexact materials category. In resource recovery, it is applied to a mixture of copper, copper alloys, zinc, lead, stainless steel, aluminum, and small amounts of precious metals. As can be seen, it is a misnomer because of the presence of stainless steel. Some authorities have suggested the term "nonmagnetic metals" for this reason. Also, the terms "other nonferrous metals" or "nonferrous metals excluding aluminum" have been used. However, in the facilities which have recovered these materials, the mixture usually has contained cast aluminum. This is due to the concentration on aluminum can recovery in most of the materials recovery technologies.

The survey shows no facilities which are operating and recovering nonferrous metals. The ten plants that were built with the capability to recover this fraction are all closed save one, Dade County, Florida. However, at present this facility does recover and sell a mixed nonferrous metal product.

The market for this mixed fraction is a secondary materials processor. This is because the constituent metals must be separated before they can be reused by industry.

Wastepaper

Paper recovery depends primarily on source separation. However, some mechanical separation of paper is likely in the future and, therefore, would affect the MSW input tonnage and composition of a recovery plant. In 1980, 12.7 million tons of paper were recovered and recycled. About two-thirds of this quantity was recovered from postconsumer waste; the remainder originated in fabricating and converting operations.

Most of the paper now received is obtained through source separation and separate collection. Some is obtained through hand sorting mixed waste, and almost none is recovered through mechanical separation from mixed waste. The paper now recovered from postconsumer sources represents a high-value (clean, high-grade paper) and readily recoverable fraction of paper discards.

The major paper industry market for postconsumer recovered paper is combination boxboard. Boxes of this material are used for packaging dry foods, shoes, clothes, and similar items. There are several grades of boxboard. A typical mix by one manufacturer requires 0.48 ton of mixed wastepaper, 0.25 ton of corrugated paper, and 0.21 ton of old newspapers to produce 1 ton of combination boxboard. The remaining 0.06 ton of material consists of the coatings. Mixed wastepaper is also used in the manufacture of building materials, roofing felt, construction paper and board. Lower standards for inclusion of contaminants in this wastepaper product usually dictate lower prices. Properly segregated corrugated containers are used to manufacture new corrugated containers, and separated newspapers are used in the manufacture of newsprint. High-grade wastepaper, e.g., printing paper, is used to manufacture new printing paper, tissue, and other products.

Paper mills are located throughout the nation; however, most of the mills that now use wastepaper are located in the Northeast and North Central regions of the nation. Exports offer a market that has grown in recent years and may provide an important outlet in selected areas of the nation.

Wastepaper prices vary by grade and region. Prices rose in the 1977 boom period but then began to decline as markets collapsed in the 1978 recession. Although severe price fluctuations are unusual, prices of wastepaper like all commodities, vary over time. Selling arrangements should take into account this variability. Specifically, contracts should be established which guarantee a minimum purchase price.

Quality requirements of most markets for wastepaper can presently be satisfied only by recovery through separation at the source or hand sorting from mixed waste. In considering a resource plant, the inclusion of source separation or hand sorting of paper in the overall recovery scheme will affect overall recovery economics. It may increase the economic viability or decrease it, depending on market prices for the recovered paper and for the energy or fuel outputs of the recovery plant.

MARKET DEVELOPMENT

With the preceding overview of products and markets in mind, the discussion will now turn to the market development plan. This discussion addresses the manner in which a market or markets can be identified and firmly committed to a project. This process is divided into three stages: identification, commitment, and contracts.

Identifying a Market

Waste-to-energy projects in small- and medium-size communities generally have a limited number of potential product markets. Because steam is typically the most valuable and salable product, with electricity a distant second, it is advisable to identify and establish contact early with industries with, ideally, two characteristics: (1) use of substantial quantities of process steam and/or hot water, and (2) internal generation of combustible wastes. For the latter, rising disposal costs may represent a substantial incentive to join a municipally sponsored waste-to-energy venture. Among the more obvious industries which meet both criteria are paper products and textile manufacturers. Industries that meet the first criterion, high process steam use, include industrial organic chemicals, fats and oils, and plastics manufacturers.

Although projects to date are dominantly single-customer industry markets, no technological barrier prevents steam and hot water sales to multiple customers involving industrial, commercial, and institutional users. Such options may be particularly attractive in cases where a waste-to-energy facility forms an early and integral part of an industrial park development. In this instance, the location of the facility and its energy products can be planned in coordination with the park's occupants, an approach which is likely to enhance the marketability of the sites under a banner of energy self-sufficiency. However, there must be a base customer that can support the economics of the project. Without a base customer, the municipality may not elect to support a more expensive disposal source, despite its future potential. This was the case in Genesee Township, Michigan where a 100 TPD modular system was built, started up, and abandoned because energy customers never located in its industrial park.

Before initial overtures are made to potential customers, the anticipated waste stream must be characterized as accurately as possible. Although existing studies of various areas of the country are useful guide points, a systematic sample of the local waste stream should be evaluated to identify any major deviations from the norm with respect to variables such as energy value, explosives, and toxicity. These samples can be compared to data compiled by the U.S. EPA and Department of the Interior's Bureau of Mines on waste stream composition. When a satisfactory estimate of waste quality and quantity is achieved and translated into expected energy value, market contacts should be initiated.

For potential energy product users, substantial information is needed in order to determine if that potential can be realized. The required data include descriptions of energy production facilities, future energy requirements, types and amounts of fuels consumed, availability of on-site space, energy demand curves, values of alternative fuel/energy sources, economics of present energy production, adaptability of integrating the use of either solid fuels or steam and electricity, overall interest in using new energy sources, and financial stability to insure product purchase under long-term contractual arrangements. An efficient means of obtaining this information is through a survey questionnaire. This form can be mailed to all potential markets and followed up with a phone call or personal visit to the most promising respondents. Once this information is gathered, an evaluation will be required to determine whether one or several installations can provide the required consumption levels for the different energy outputs.

The following criteria should be used to select entities for further consideration as markets for recovered energy:

- *Technical feasibility.* The use of recovered energy by the market is not precluded by technical constraints.
- *Energy demand.* The energy demand of the market is sufficient to consume all or a significant amount of the solid waste generated in the landfill service area where it is located.
- *Load uniformity.* The seasonal variation in the energy demand of the

market is insufficient to cause a significant loss of revenues for energy sales from a resource recovery facility.

- *Location.* The market is located sufficiently near the waste generation centers to prevent high transportation costs but is away from residential areas.

Users of the various material products in the waste stream are often more familiar than energy users with the use of the recovered product in their industry. Certain industries such as paper, steel, aluminum, and glass have established specifications for purchasing secondary products and have experience with using such products in their manufacturing process. For these reasons, it may be less difficult to assess material product users' interest in the recovered product. A sample survey of potential markets and their characteristics for a proposed Salt Lake City facility appears in Table 4-7.

Once the preferred users have been identified, future activities should be closely coordinated to ensure their continuing and active role in future project planning. This market role should not be underestimated. Various project development experiences suggest that markets are hardly passive. In the North Little Rock, Arkansas modular incinerator facility, for example, the future energy user took an active role in the development and implementation of the project. This was also the case in Auburn, Maine, and Windham, Connecticut, where industry involvement was instrumental throughout the project development process.

Commitment Stage

With potential markets identified, a closer look at the processing technologies available to meet market specifications should come next. Technological risks, system capital and operating costs, alternative facility sites, overall system logistics and costs, financial and management alternatives, required legislation/authority formation, and budget requests should be evaluated. As this other information is developed, a further refinement of market development should take place prior to any decision on project "go/no-go." The development of more detailed documents such as "Memoranda of Understanding," "Letters of Intent to Bid for the Purchase of Recovered Products," or actual "Invitation to Bid for Recovered Products" should then be undertaken.

"Letters of Interest," or preferably "Letters of Intent," are essentially advance contractual purchase commitments. The objective of the advance commitments is to go beyond a general statement of willingness to consider the future purchase of recovered products if and when they become available. Financial planning for resource recovery, on either a public or a

Table 4-7. Potential Energy Markets Identified for a Proposed Salt Lake Waste-to-Energy Facility.

POTENTIAL MARKET	OPERATIONS	SEASONAL DEMAND	DEMAND FOR STEAM OR OTHER NON-ELECTRICAL ENERGY	STEAM CHARACTERISTICS	FUELS USED	ANNUAL ELECTRIC POWER DEMAND	AVAILABILITY OF SITE FOR ENERGY RECOVERY FACILITY	LEVEL OF INTEREST	POSSIBILITY OF LONG-TERM CONTRACT
Billings Clinic	8 hr/day 5.5 days/wk	No	8,607 cfm/yr	15 psi—140°F	Natural gas	1,356,480 kWh	—	—	—
Billings Deaconess Hospital	24 hr/day 365 days/wk	No	5,000 lb/hr (80%) 2,000 lb/hr (20%)	60 psi—300°F 30 psi—271°F	Natural gas	2,714,400 kWh	No	High	Yes
Midland Foods	8 hr/day 5 days/wk	Yes	300,000 lb/hr	120 psi—360°F	Natural gas	8,130,94 kWh	Yes	Medium	Yes
Rocky Mountain College	24 hr/day 365 days/wk	Yes	—	—	Natural gas	—	No	Low	No
Roscoe Steel & Culvert Co.	16 hr/day 5 days/wk	No	—	—	Natural gas	—	Yes	Medium	—
St. Vincent's Hospital	24 hr/day 7 days/wk	No	30 lb (85%) 40 lb (15%)	—	Natural gas	6,213,600 kWh	No	High	Yes
The Lovell Clay Products Company	24 hr/day 7 days/wk	No	90,172 mcf/yr	—	Natural gas	408,000 kWh	Yes	Medium	Yes

private basis, must be based on a well-defined revenue stream. In this way, some of the economic uncertainty is removed as a trade-off against technical uncertainty.

Letters of Intent (LOIs) obtained for products to be recovered from a publicly owned facility must be structured differently than for a privately owned facility in order to be within the general legal framework of public bidding laws. As such, the LOI for a publicly owned facility becomes a "Letter of Intent to Bid for the Purchase of the Recovered Product." Later, the competitive bidding process can take place for the actual sale of the recovered product, or the LOIs obtained as part of the initial feasibility analysis can be assigned to a private organization if the recovery is to be done by this mode.

An LOI should include the following elements:

- *Delineation of product.* The LOI should contain specifications for the energy product to be sold.
- *Quantities.* The quantity of the energy product to be delivered should be specified. This could involve a "put-or-pay" clause in the LOI.
- *Delivery schedule.* A delivery schedule for the specified quantity of the energy product should be agreed upon in the LOI. This is particularly critical for steam and electricity.
- *Duration.* The duration of an LOI agreement should cover the planning and procurement stages of a resource recovery facility. The LOI agreement for energy products should last the planned life of the facility (usually 20 years) since the revenues for energy are critical for the economic viability of the project.
- *Specification.* The LOI should agree on specifications for the recovered energy product which are reasonable for both the user and the seller.
- *Pricing structure.* The pricing should allow for escalation in relation to alternative fuels, sources of energy, or general inflation.
- *Assignability.* The LOI should be assignable in case the actual operation of the resource recovery facility is performed by an entity other than the one involved in negotiating the LOI.
- *"Out" clauses.* In the event that implementation of a resource recovery project is delayed, the LOI should not bind the purchaser of energy. A clause should be included allowing termination of the LOI if substantial progress toward implementation is not made.

These elements are specifically applicable to energy products; however, the principles also apply to materials, as will be discussed.

In negotiating prices for energy products (steam, solid fuels, and electricity), the goal is to establish a fair and equitable price for both the producer and the user. This price should provide an incentive to produce the fuel as well as to use it. Hence, the price should lie somewhere between the net cost of production and the net value of the fuel to the user, both of which are subject to the uncertainties of any long-term forecast. Also present is the problem of apportionment in applying the operating and maintenance costs of the various items when there is more than one product.

Contract Stage

For material products, the "commitment" document should include terms and conditions with respect to floor prices and exchange price formulas, length of commitment, quantity, quality, delivery schedule, and termination provisions. From the seller's standpoint of course, long periods over which materials are purchased are desirable. Five year periods are often specified, especially in Letters of Intent. This is adequate, although a ten year period would be preferable, since it would be more representative of a project's amortization and operational lifespan. This is in contrast to the energy markets in which a 20 year agreement is the norm.

As the project development process proceeds, commitment agreements which sufficed for purposes of financial planning must be transformed into formal, legally binding contracts. Alternative contractual arrangements utilized by representative projects are the subject of detailed discussion in Chapter 9.

CONCLUSION

The central importance of market guarantees in implementing a resource recovery system cannot be overstated. Early identification of potential energy and materials customers—and involvement of the same throughout the project development process—are cornerstones to a successful project. To date, industrial steam has proven to be the most marketable energy product produced by waste-to-energy facilities. Industries which both consume large quantities of process steam and internally generate large volumes of combustible wastes are the most promising candidates under current conditions. However, under recent federal regulations, cogeneration of steam and electricity, or electricity alone, will become an increasingly viable option for developers of small- and medium-size projects. Recovered materials, on the other hand, will remain of relatively minor financial significance owing to the limited volume, high cost, and spotty technical experience of separation techniques. Thus, in the foreseeable future, project sponsors that target proven industrial steam markets such as textile, paper, and chemical producers, in either single or multiple customer contractual arrangements, and/or electricity sales to electric utilities are likely to achieve early successes in developing waste-to-energy projects in small- and medium-size ranges.

5
SMALL-SCALE TECHNOLOGY

A review of solid waste energy recovery technologies is presented in this chapter. Particular emphasis is placed on technologies commercially available for converting solid waste to energy. These technologies are:

Mass Burn
- Direct combustion (field erected)
    Waterwall incinerators
    Refractory-lined incinerators
- Modular incinerators (shop fabricated)

Refuse-Derived Fuel (RDF)
- Dry processing
- Wet processing
- RDF co-fired with conventional fuels
- RDF dedicated boiler

Under each of these technologies, information is given on the combustion process, background of the technology, current status, principal vendors, and problems experienced.

MASS BURNING

The direct incineration of solid waste with energy recovery can be categorized into two basic technologies: direct combustion field-erected processes and modular units that are mainly shop fabricated. With the exception of removing large, bulky items such as white goods and mattresses, no intermediate processing of the refuse is required with these mass burning units. Any large items that have to be removed from the refuse stream, mainly because of physical size constraints, must go to a landfill or possibly salvage. All other combustible waste materials (e.g., paper, plastic, wood), as well as noncombustible items (e.g., glass, metal, grit, etc.), are introduced into the mass burning system's feed chute and, eventually, to the firebox where the combustion of fuel takes place. At some of the large-scale mass burning facilities [e.g., Pinellas Co., Florida, 2,000 TPD (tons per day) capacity], significant amounts of residue allow back-end separation. Ferrous metals and aggregate are separated from the furnace ash residue prior to disposal at the landfill.

Direct combustion mass burning units are designed for the total combustion of refuse in one large primary combustion chamber, whereas dual chamber modular mass burning units use both a primary combustion chamber and a secondary chamber for fuel combustion and volatile gas destruction.

Direct Combustion

Direct combustion is the traditional technology for incinerating municipal solid waste (MSW). Many facilities with long-term operating experience are located in European and Asian countries where both landfill and energy scarcities expedited development. Direct combustion technology is separated into two subsystems: waterwall incinerators in which the furnace (combustion chamber) and boiler are integral components, and refractory-lined furnaces with a convection-type waste heat boiler located downstream from the furnace chamber.

Waterwall Incinerators.   Waterwall incinerators operate in essentially the same manner regardless of their capacity. All employ the use of water tubes in the furnace area that contains circulating water. Figure 5–1 depicts a typical waterwall incinerator. The moving liquid serves to dissipate the high heat associated with the combustion process and thus reduces the amount of cooling air required. In addition, the circulating boiler water also serves as a heat recovery medium, aiding in the generation of steam for energy recovery systems.

In typical waterwall systems, refuse is transferred from the refuse storage pit by an overhead bridge crane, fitted with a grapple, to the refuse feed hopper/chute assembly. The hopper/chute assembly is kept full of solid waste. By gravity flow, it leads directly to the feed rams at the top of the combustion chamber stoker assembly. The MSW feed chute may be water-cooled or refractory lined to protect exposed metal parts, due to the fact that some refuse may smolder and burn in the chute, causing high interfacial temperatures between the refuse and chute. By keeping the chute water-cooled, heat deformation of the metal is avoided. At the feed chute/stoker interface, the refuse may be either gravity fed onto the stoker or more positively fed by a hydraulic ram feeder unit. Within the combustion chamber, the stoker assembly extends from the end of the feed chute to the ash (residue) discharge area. The stoker consists of a number of grate levels and individual grate bars that support the combustion or refuse by permitting the entry of undergrate air (i.e., underfire air) into the combustion

Typical Simplified Boiler Cross Section

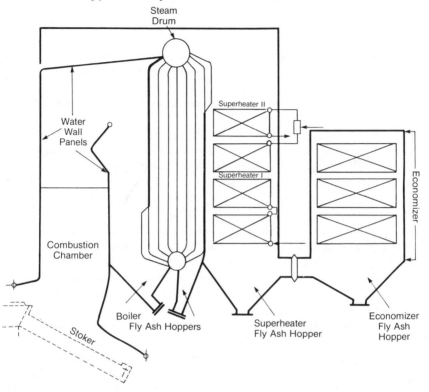

Figure 5-1.   Cross-section of Mass-Burning waterwall incineration facility. *(Courtesy of Ogden-Martin Systems Inc.)*

chamber. Depending on the manufacturer, the stoker provides the mechanism for the tumbling and mixing of refuse by either mechanical or hydraulic action. The various types of stoker assemblies are depicted in Figure 5-2. Of those shown, the reciprocating stoker is the most widely used in operating incinerators.

One waterwall system stoker manufacturer, the O'Connor Combustor Corporation, acquired by Westinghouse in late 1983, employs a system for the incineration of refuse that is quite different than the conventional grate system. The O'Connor system employs a device similar to a chemical plant kiln that is lined with water tubes. Refuse is ram fed into the circular water-walled combustor through a traditional gravity fed chute and hydraulic rams. The combustor cylinder has a slight downward incline (i.e., a 6° slope) and slowly rotates about an imaginary horizontal axis. This arrangement allows for the tumbling, mixing, and aeration of the refuse, which aids in its complete incineration. The water tubes of the combustor tie into a common header at the downward end of the combustor, which in turn ties into the tube sheet of a waterwall boiler. About 30 to 35 percent of the energy is generated in the combustor, the rest in the boiler. Combustion air is supplied through louvers on the underside of the rotary combustor unit. Six air zones admit preheated combustion air through the holes in the web. Ash is discharged at the end of the unit into an ash quench–type disposal system (see Figure 5-3).

Typically, within any of the combustion chamber designs there are three zones of activity: (1) a drying zone, (2) a combustion zone, and (3) a burn-out zone. The drying zone is needed to drive off the moisture from the refuse so that when the refuse enters the combustion and burn-out zones, it will burn completely and efficiently. The air required for the combustion of refuse is supplied by the forced draft (FD) fan. Ductwork from the FD fan distributes primary combustion air to air plenums located beneath the stoker grates, as well as to secondary air injectors located above the grate system around the incinerator firebox. The injection of secondary air allows for the control of the fireball height and ensures that any gaseous products of combustion and burning refuse particles, or "sparklers," which may rise from the stoker bed, are completely burned. Waterwall incinerators usually require excess air in the range of 40 to 80 percent above stoichiometric conditions.[1]

Because the boiler unit of most waterwall systems is usually installed above the furnace (i.e., combustion chamber), most refuse-fired waterwall units have upper limits for steam conditions in the range of 650 psig and 750°F. This is typically set to avoid high-temperature corrosion of the waterwall tubes caused by the waste stream constituents. Pressures and temperatures above these settings require that incrementally thicker boiler tubes or expensive alloys be used to provide a corrosion allowance, as well as to ensure boiler integrity. Some large-scale units, such as the Signal-RESCO facility at Saugus, Massachusetts, have been designed to operate above these conditions and can do so because of the large-scale project economics involved. With superheater steam generated for power production applications, the temperature is controlled by attemperators that inject water or steam into the superheater header to control abnormally high-temperature steam. The superheater may be either a single pass or a multiple pass unit, depending on the degree of superheat required for the steam. Multiple pass

---

1. Stoichiometric condition refers to the amount of air containing the *exact* amount of oxygen required for the full and complete combustion of refuse.

RECIPROCATING GRATES

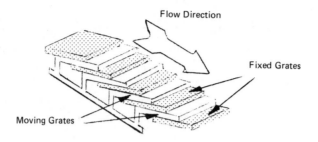

ROCKING GRATES

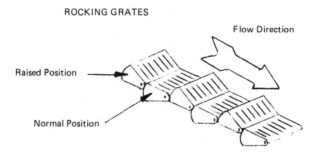

TRAVELING GRATES

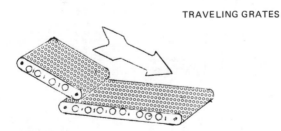

Figure 5–2. Typical incinerator grate assemblies.

Westinghouse/O'Connor
Resource Recovery Plant

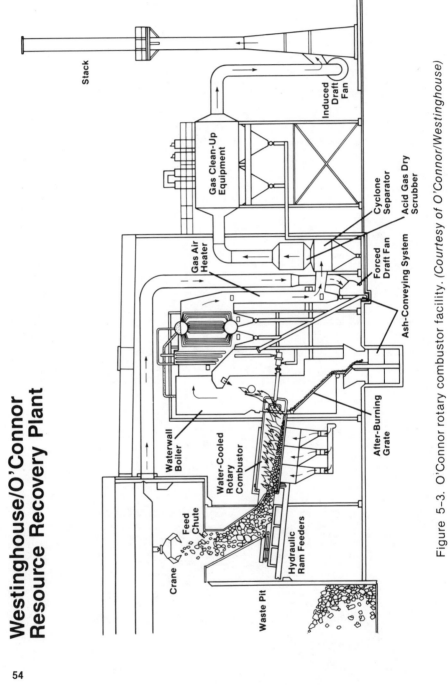

Figure 5–3. O'Connor rotary combustor facility. *(Courtesy of O'Connor/Westinghouse)*

units, because more area is available for heat transfer, provide for a greater amount of superheat but are also more expensive.

The waterwall tubes tie into common headers at the base and at the top of the boiler. The upper waterwall header is suspended from the ceiling or roof of the facility, while the bottom header is supported by a high-tension spring suspension system. The superheater tubes also tie into a common header and, like the waterwall tubes, hang vertically inside the boiler. This type of floating construction allows the boiler tubes to expand and contract, with minimal stress to the unit, when temperature variations occur inside the boiler. This will occur several times per year as the units are taken off-line for maintenance activities. Some units may be fitted with an economizer, or economizing bank, which is situated downstream of the boiler itself. The economizer uses waste heat (in the form of hot flue gas after the boiler) to heat boiler water. Structurally, the economizing section is a bank of tubes that horizontally span the width of the flue gas duct. It is situated downstream of the boiler and upstream of the air pollution control unit.

Water to the boiler is supplied from either returned steam condensate or boiler makeup water. The introduction of untreated water directly into the boiler tubes is not permissible. The untreated water must be purged of solids and any harmful chemicals that may cause boiler tube scale or plugging. After proper treatment, makeup water combines with condensate in the deaerator tank. Here, all gases are expelled from the boiler water. It should be noted that this is an essential step in the steam generation process as any oxygen in pressurized water is very corrosive to metals. Thus, failure in removing the oxygen could lead to the water-side corrosion of the boiler tubes. From the deaerator tank, water is drawn to the feedwater pump where it is pressurized and transferred to the boiler. A multistaged centrifugal pump is needed to generate the proper pressure head required to transfer the pressurized water and generate steam in the boiler.

Bottom ash is discharged at the end of the stoker into an ash handling system. This system could be of the dry type or the wet/quench type. A dry ash removal system allows the ash to fall into a large pit or hopper. At this point, it is either manually or mechanically removed from the pit. A wet/quench unit, which is the common approach used with mass burning systems, allows ash to fall into a water-filled trough where a drag conveyor dredges the bottom of the pit for ash. Advantages in deploying a wet/quench unit are:

- The wet system provides for an air seal of the furnace.
- The ash is kept in solution with water.

The former prevents any gaseous effluent from leaving the furnace and prevents any tramp air from entering the unit, which could cause poor flame and combustion patterns, as well as other operational problems. The latter prevents the ash from entering the atmosphere (characteristic of a dry system), thus eliminating a serious health hazard. Gaseous effluent is removed from the combustion chamber and through the boiler by the induced draft (ID) fan, and then treated by an air pollution control device. Depending upon the project economics, regulatory requirements, and size of the project, either a baghouse or an electrostatic precipitator would typically be used. Depending mainly on state regulatory requirements, additional controls in the form of acid gas scrubbing may be required. This additional level of gaseous emission control would add substantially to the project costs.

Depending on the amount of air needed for proper combustion and furnace temperature/pressure control, the waterwall direct combustion system can recover energy with efficiencies ranging from 65 to 75 percent. At burning rates lower than 50 percent of nameplate capacity, lower efficiencies and various operating problems can be expected. This is due to a nonuniform distribution of fuel on the grate which causes air to circumvent fuel through bare spots in the fuel bed. Combustion air will follow the path of least resistance (e.g., it is much easier for air to flow through a hole in a grate bar than through a foot or more of refuse). The inadequate fuel/combustion air mixing causes the incomplete firing of refuse in the furnace, thereby affecting combustion rates and efficiencies.

Refractory-Lined Incinerators.   Refractory-lined refers to a thick (6 to 8 inch), heat resistant coating in the furnace that both decreases the transfer of heat produced from the combustion process to areas outside the incinerator unit, and protects the outer metal shell from extreme and sudden changes in furnace temperature (refer to Figure 5-4). As can be seen from Figure 5-4, the main difference between a refractory-lined unit and a waterwall unit is that with a waterwall unit, the boiler is part of the combustion chamber, whereas in refractory-lined units, the boiler is located downstream of the combustion chamber.

Like the waterwall system, the refractory-lined unit is fed by a bridge crane fitted with a grapple. The refuse falls by gravity through the feed hopper/chute assembly to the feed table. As with the waterwall unit, the feed chute may be water-cooled. Refuse is usually fed to the stoker assembly by a ram feeder, or it falls by gravity from the edge of the feed chute onto the stoker. The stoker provides the means by which refuse is combusted in the incinerator. Depending on the incinerator vendor, the grate assembly will be one of those previously mentioned.

Energy recovery from refractory-lined units is facilitated by the installation of a waste heat boiler positioned downsteam of the furnace section. The hot gases from combustion are removed from the furnace by the induced draft fan and passed through the waste heat boiler. The gases release their

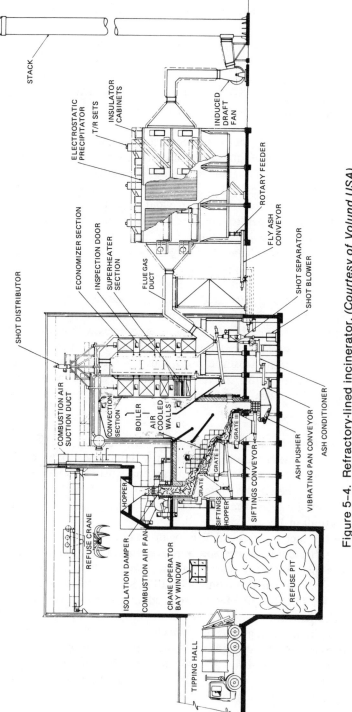

Figure 5-4. Refractory-lined incinerator. (*Courtesy of Volund USA*)

STACK

ELECTROSTATIC PRECIPITATOR

T/R SETS

INSULATOR CABINETS

INDUCED DRAFT FAN

ROTARY FEEDER

FLY ASH CONVEYOR

SHOT SEPARATOR

SHOT BLOWER

ECONOMIZER SECTION

INSPECTION DOOR

SUPERHEATER SECTION

FLUE GAS DUCT

SHOT DISTRIBUTOR

COMBUSTION AIR SUCTION DUCT

CONVECTION SECTION

BOILER

AIR COOLED WALLS

ASH CONDITIONER

ASH PUSHER

VIBRATING PAN CONVEYOR

SIFTINGS CONVEYOR

GRATE III

GRATE II

GRATE I

SIFTINGS HOPPER

HOPPER

REFUSE CRANE

ISOLATION DAMPER

COMBUSTION AIR FAN

CRANE OPERATOR BAY WINDOW

REFUSE PIT

TIPPING HALL

57

heat to the boiler tubes by convection, and through the conductive heat transfer phenomenon, the boiler tubes transfer this heat to the pressurized boiler water. Upstream from the convection or boiler section of the unit, a superheater section may be located. Downstream from the boiler, an economizing section may be situated to use waste gas heat to increase the overall system efficiency.

Because the refractory unit is not equipped with a radiant boiler section (i.e., water tube exposed to the combustion fireball) as is the waterwall unit, the boiler efficiency with a refractory-lined incinerator will be less than that of a waterwall system. The efficiency difference may be on the order of 3 to 10 percent, depending upon facility size and system manufacturer.

Refractory-lined incinerators require a large volume of air (up to 200 percent excess) to provide both combustion air for the burning process and cooling air for the furnace wall and the feed grate system. Since the furnace draft system (FD fan, ID fan, appropriate ductwork, and motors) and the air pollution control equipment component costs are directly related to the quantity of air to be used/treated, these incremental component costs are higher with refractory-lined units than with waterwall incinerators. Residue from the grates is disposed of in the same manner as with a waterwall unit. The flue gas, because of the large volume of air that must be processed, usually requires an electrostatic precipitator for treatment. It should be noted that only one existing conventional refractory-lined incinerator in the United States has been retrofitted with waste heat boilers (Waukesha, Wisconsin).

Background: Waterwall Incinerators.   During the mid-1950s, several European companies began to experiment with methods to improve energy recovery from MSW. The costs of energy and waste disposal, which were higher in Europe than in the United States, provided the incentive for this experimentation. In 1956, the first waste-to-energy plant with a watertube boiler was built in Bern, Switzerland.

Despite the fact that several European combustion systems were being marketed in the United States and were recognized as the state of the art in solid waste incineration, the first waterwall system installed in the United States was a unit designed and built with all U.S. manufactured components. In 1967, a 360 TPD two line facility to produce process energy began operations at the Naval shipyard in Norfolk, Virginia. This plant employed Detroit Stoker grates and Foster Wheeler boilers. However, the facility failed to represent the opening of a widespread market. Since it was a government procured, owned, and operated system with a relatively specialized waste stream, this type of facility generally was viewed as having limited application to the municipal market.

In addition to the potential for higher-quality energy, waterwall incinerators had another advantage that would help to penetrate the municipal market. Less combustion air is necessary with waterwall incinerators, since the water tubes help to cool the walls of the combustion chamber. A larger volume of combustion air is needed to cool the walls of a refractory-lined incinerator. This is an important consideration following enactment of the Clean Air Act in 1970 as air pollution control equipment costs are directly related to the amount of stack gas that must be treated.

As shown in Table 5-1, four additional small-scale waterwall incinerator systems that use MSW have been constructed, and are in operation, in the United States since the completion of the Norfolk facility in 1967. In addi-

Table 5-1. Waste-to-Energy Systems: Small-Scale Waterwall Incinerators.

YEAR OF START-UP	LOCATION	DESIGN CAPACITY (TPD)	ENERGY MARKET
1967	Norfolk, VA	360	U.S. Naval Base Norfolk
1971	Braintree, MA	250	Weymouth Art Leather Co.
1980	Hampton, VA	200	National Aeronautics and Space Administration
1982[a]	Gallatin, TN	200	Industrial park complex and TVA
1984[a]	New Hanover Co., NC	200	W. R. Grace (steam) and Carolina Power and Light (electricity)
PLANTS UNDER CONSTRUCTION			
1987[a]	Marion Co., OR	550	Portland General Electric Co. (electricity)
1987[a]	Tulsa, OK	750	Sun Oil Co. (cogeneration)
1987[a]	Dutchess Co., NY	400	IBM (steam), Central Hudson Gas & Electric (electricity)
1987[a]	Olmsted Co., MN	200	Olmsted Co. (steam & electricity), Federal Bureau of Prisons (steam & electricity) and public utility (electricity)
1987[a]	Commerce City, CA	300	All electric
1987[a]	Bay Co., FL	510	All electric
PLANTS UNDER FINAL CONTRACT NEGOTIATIONS WITH SELECTED VENDORS			
1987[a]	Claremont, NH	230	Central Vermont Public Service Corp. (electricity)
1987[a]	Islip, NY	520	Long Island Lighting Co.

[a] Electric generation or cogeneration.

tion to steam, electricity is also generated by two of these facilities: (1) Gallatin, Tennessee and (2) New Hanover County, North Carolina. At the facility in Gallatin which began operations in the spring of 1982, up to 550 kilowatts of electricity is produced. The New Hanover County, N.C. power sales contract is with Carolina Power and Light, and the County will generate approximately 2,250 kilowatts and will sell electricity directly to the electrical power grid operated by the local utility.

During the 25 year period from 1958 to 1982, five European vendors built 79 plants in Asia and Europe that involve electrical generation from waste-to-energy plants. Some of the facilities produce only electricity and the remainder are cogeneration plants. Pertinent data regarding a few of these plants are given in Table 5-2. Additionally, a Japanese company, Takuma Ltd. (known for its boiler making operations), licenses its own reciprocating grate system and has constructed over 230 waste incineration facilities in Japan. Of these facilities, 15 plants produce steam or hot water through waterwall boilers, and 18 produce both steam and electricity (cogeneration) through waterwall boilers and turbine generator sets. Their largest unit is a 2,040 TPD cogeneration plant producing 15 megawatts of electricity.

Background: Refractory-Lined Incinerators.   Direct combustion, or incineration, of MSW in a facility designed specifically for that purpose has been practiced in the United States since 1883. Refractory-lined units were the only type of combustion system associated with solid waste incineration until the 1950s. The first reported use of MSW to produce energy took place in Hamburg, West Germany in 1886. In the United States, recovery of the energy from MSW began in 1898 with a direct combustion system. This facility, which was located in New York City, used a waste heat boiler to recover the heat of combustion.

Actual combustion of the MSW occurred in a refractory-lined incinerator, whose primary purpose was waste reduction, rather than energy recovery. As with most other such systems that followed, the steam produced was used only for the process needs of the facility. However, in 1903, electricity was being generated in at least one facility that also was located in New York City.

Use of MSW as an energy resource in this country declined in the early part of this century due to the advent of lower-cost fuels. In Europe, where the cost of disposal and energy has been relatively higher than in the United States, energy recovery has been a more important part of incineration. The majority of system vendors for refractory units as well as waterwalls are representatives or hold licenses from European companies.

Even though energy recovery was an unimportant part of incineration in the United States, the number of refractory units, which were used to reduce

Table 5-2. Representative Waterwall Facilities in Asia and Europe.[a]

LOCATION	NUMBER OF UNITS[b]	UNIT RATED SOLID WASTE CAPACITY (TPD)[b]	START-UP YEAR ELECTRIC PRODUCTION[b]
Biebeshein, W. Germany	2	75	1980
Bremerhaven, W. Germany	3	240	1977
Frankfurt am Main I, W. Germany	2	360	1966
Frankfurt am Main II, W. Germany	2	360	1967
Hamburg II, W. Germany	3	200	1963
Landshut, W. Germany	1/1/1	72/72/144	1971/1974/1981
Ludwigshafen Stadt, W. Germany	2	240	1967
Nuremberg, W. Germany	3	360	1968
Soliñgen, W. Germany	2	240	1970
Helsinki, Finland	2	200	1961
Dijon, France	2	300	1974
Strasbourg, France	3	336	1974
Den Haag, Netherlands	3/1	360	1967/1974
Genoa, Italy	3	240	1972
Chiba, Japan	3	150	1977
Higashi Osaka, Japan	2	300	1981
Kitakyushu, Japan	2	300	1977
Nagano, Japan	3	150	1982
Osaka Minato, Japan	2	360	1977
Osaka Nishiyodo, Japan	2	200	1964
Osaka Taisho, Japan	2	360	1980
Sendai, Japan	3	200	1976
Suita, Japan	3	165	1981
Tokyo Adachi, Japan	4	316	1977
Tokyo Ohi, Japan	4	300	1973
Tokyo Setagaya, Japan	3	300	1968
Tokyo Shakujii, Japan	2	300	1968
Toyonaka-Itai, Japan	3	225	1975
Yokohama, Japan	3	420	1980
Vienna II, Austria	2	506	1980
Vienna I, Austria	3	200	1963
Norrtorp, Sweden	1	110	1982

[a]Source: Gershman, Brickner & Bratton, Inc.
[b]Multiple listings signify expansions of original plants.

the volume of MSW, continued to expand. Prior to 1970, when the Clean Air Act (CAA) became law, there were 287 refractory-lined incineration plants in operation. None of these plants were reported to be generating electricity and only 32 recovered energy. Most of these plants used the energy for in-plant requirements only.

Following the enactment of the Clean Air Act in 1970, about 50 percent of the existing conventional solid waste incinerators closed. The cost of the air pollution control equipment necessary to comply with the CAA regulations made these incinerators—particularly those using old, inefficient units—uneconomical. One pre-CAA system that continues to operate is the Betts Avenue incinerator in New York City. Process energy for the incinerator and for adjacent City buildings is produced by this facility.

Since 1970, three conventional field erected refractory-lined incinerators (with energy recovery) have begun operations: (1) Waukesha, Wisconsin; (2) Glen Cove, New York; and (3) Harrisonburg, Virginia (see Table 5–3). The Waukesha plant was an existing refractory incinerator built in 1971 that was retrofitted with a waste heat boiler in 1979. Glen Cove is a new codisposal (MSW and sludge) plant which commenced commercial operations in August 1983. Both facilities generate electricity. The Harrisonburg facility began operations in November 1982. At present, steam from the Harrisonburg facility is only used for district heating by James Madison University. Several options are being considered for steam use during the summer, including electrical generation.

Table 5–3. Waste-to-Energy Systems: Refractory-Lined Incinerators.

YEAR OF START-UP	LOCATION	DESIGN CAPACITY (TPD)	ENERGY MARKET
1965	New York, NY (Betts Avenue)	500	District steam loop
1971[a] (1979)[b]	Waukesha, WI	175	Amron Corp. (cogeneration)
1982	Harrisonburg, VA	100	James Madison University
1983[a]	Glen Cove, NY	250	Sewage treatment (electricity)
PLANTS UNDER CONSTRUCTION			
1985[a]	Tampa, FL	1,000	Tampa Electric Company (electricity)
1985	Susanville, CA	100	Lassen Community College
1986	Key West, FL	150	—
PLANTS UNDER FINAL CONTRACT NEGOTIATION WITH SELECTED VENDORS			
1987[a]	Davis Co., UT	500	Hill Air Force Base (cogeneration)
1987[a]	Warren/Washington Co., NY	400	—
1987	Savannah, GA	450	American Cyanamid

[a] Electric generation or cogeneration
[b] Retrofitted for electric generation.

Current Status: Waterwall Incinerators. As noted in Table 5-1, there are five smaller-scale waterwall incinerators under construction in the United States: Marion County, Oregon; Tulsa, Oklahoma; Commerce City, California; Olmstead County, Minnesota; and Bay County, Florida. All of these smaller-size facilities are designed for electrical generation, while two of the projects will include cogeneration with steam sales to local industries.

Two other smaller projects have selected the waterwall design system over other technologies to convert their waste to energy. One of these projects will produce electricity while the other will be a cogeneration plant.

Current Status: Refractory-Lined Incinerators. At present, refractory-lined incinerators are under construction in Susanville, California; Tampa, Florida; and Key West, Florida (see Table 5-3). The Susanville facility is designed to produce process steam for a district steam loop, while the larger Tampa project will generate 25 megawatts of electricity for Tampa Electric Co. when it comes on-line.

In addition, three other communities have selected system vendors of refractory-lined incinerators with waste heat boilers as their energy recovery process. These facilities are to be built in Warren/Washington County, New York; Savannah, Georgia; and Davis County, Utah. Electricity will be generated by the first facility and the other projects will involve the cogeneration of electricity.

Future Developments. A potentially important development with waterwall system designs is the evolution of modular components for small-scale (400 TPD or less) systems. Modularization is expected to lower on-site construction costs, thus reducing the capital cost of small-scale waterwalls and the time required to build the plants. From a power generation perspective, increased use of waterwalls is preferable because of the higher temperature and pressure that can be produced as well as the higher energy efficiency levels. Higher pressure enables the steam to have a greater pressure drop (i.e., do more work) through a turbine, which increases the power output.

Three factors could potentially influence the use of refractory-lined incinerators for energy recovery systems. First, if the Harrisonburg, Virginia and Glen Cove, New York facilities and the others scheduled to begin operations in the next few years are technically reliable and cost effective, there will be increased interest in this process. One option for increasing cost effectiveness of smaller systems (400 TPD or less) is to modularize some components to decrease field erection time and costs. This is being done with waterwall systems, but the design size of the furnace and appropriate temperature and air flow patterns for good combustion may limit the downsizing of these units. No known efforts are underway with refractory units, however. Another option to increase the use of refractory-lined incinerators for energy recovery is the retrofitting of existing incinerators with a waste heat boiler. Many closed incinerators might currently be inoperable, but the system might be suitable for retrofitting with energy recovery, new air pollution control systems, and other field modifications. This still may be more cost effective than starting from scratch.

Vendors: Waterwall Incinerators. The major small-scale system vendors that are active in the waterwall area are listed in Table 5-4. As with the refractory-lined incinerators, the European system vendors predominate.

Clark-Kenith, Inc. (CKI) is a new company that resulted from the purchase of Kenith Energy Systems by Clark Enterprises, Inc. J. M. Kenith Co. was the construction company that built the 200 TPD system in Hampton, Virginia. U.S. manufactured components (grates—Detroit Stoker; boiler— E. Keeler) are used by CKI. New Hanover County, North Carolina selected CKI to build a 200 TPD cogeneration system. Alexandria-Arlington County, Virginia has selected CKI to build a 975 TPD facility. The NH/VT Project (Claremont, New Hampshire area) has also contracted with CKI for a 230 TPD electricity project. In order to reduce on-site construction costs and, thus, be competitive with the modular incinerator companies, CKI is

Table 5-4. Waterwall Incinerator Vendors in the United States.

VENDOR	PROCESS
Clark-Kenith, Inc. Atlanta, GA	Detroit Stoker/E. Keeler Co. (U.S.)
O'Connor Combustor Corp. Costa Mesa, CA (Westinghouse)	O'Connor Rotary (U.S.)
Worcester, MA Riley/Takuma Ogden-Martin[a] Paramus, NJ	Takuma (Japan) Martin (Germany)
Signal-RESCO[a] Des Plaines, IL	Von Roll (Switzerland)

[a] In 1982, Wheelabrator Frye, Inc. and the Signal Companies merged. The resulting company which is responsible for waste-to-energy activities in the United States is known as Signal-RESCO. Because UOP, which is also a Signal company, marketed the Martin waste-to-energy process, the Securities and Exchange Commission ruled that UOP must relinquish its rights to the Martin system. These rights were acquired by the Ogden Corporation. As of spring 1983, the Ogden Corporation has marketed its system through its wholly owned subsidiary Ogden Martin Systems, Inc.

using new modular waterwall combustion and boiler components in the New Hanover County project. Reduced on-site construction costs, if realized, may decrease the total capital costs and thus increase the marketability of waterwall technology in small-scale applications.

The O'Connor system, acquired by Westinghouse in late 1983, uses a rotating combustion chamber that is lined with water. A 510 TPD plant in Bay County, Florida and the 400 TPD plant in Dutchess County, New York are now under construction. West Contra Costa County, California; Los Angeles, California; and Dutchess County, New York have also selected the O'Connor technology. The two New York projects are being built by Pennsylvania Engineering, Inc., acting as the full-service vendor using O'Connor equipment. In addition, there are several O'Connor units in operation in Japan, as well as two 100 TPD units at the Sumner County Facility in Gallatin, Tennessee.

In September 1984, Riley Stoker Corporation signed an agreement with C. Itoh & Co. (America) granting Riley exclusive rights to the Takuma mass burning technology in resource recovery facilities utilizing MSW in the United States. C. Itoh is a large Japanese holding company that had, as of mid-1983, been marketing the Takuma waste incineration/energy recovery system in the United States. As previously stated, Takuma has constructed over 230 refuse incineration facilities in Japan (containing 474 furnace units), 33 of which are equipped for energy recovery.

The Ogden-Martin (O-M) Corporation markets the Martin waste-to-energy system process. Martin has 189 incinerator units at 99 plants, of which 135 units include energy recovery modules. Three Martin facilities were built in this country prior to its license arrangement with UOP and, subsequently, Ogden. These facilities were large plants in Chicago, Illinois; Harrisburg, Pennsylvania; and Pinellas County, Florida. In early 1984, Ogden-Martin signed a contract to design, build, own, and operate a smaller 750 TPD cogeneration system in Tulsa, Oklahoma. In late 1984, the smaller-sized Marion Co., Oregon, 550 TPD, all-electric plant was financed. In the spring of 1985, O-M proposed on a 400 TPD system for Portland, Maine but was not selected as the preferred vendor.

Signal-RESCO (S-R) markets the Von Roll process that WFI acquired in 1971. Ninety-nine plants employ Von Roll technology around the world; forty-two are energy recovery plants in Europe and Asia. In the United States, Signal-RESCO has large facilities in Saugus, Massachusetts; Westchester County, New York; Baltimore, Maryland; Pinellas County, Florida; and North Andover, Massachusetts. The Pinellas County facility and the North Andover plant are Martin technology (UOP) plants that Signal kept the rights to when it merged with WFI. In late 1984, S-R proposed on a 500 TPD facility and has expressed interest in certain smaller-scale systems.

Vendors: Refractory-Lined Incinerators.   The leading system vendors in this technology are the U.S. representatives of European companies. The primary vendors are given in Table 5–5.

Bruun & Sorensen (B&S) of Aarhus, Denmark has been marketing directly in the United States for several years. At present, Bruun & Sorensen has a contract to build a 100 TPD plant in Susanville, California that will cogenerate steam and electricity. Recently, an agreement was reached with Lahontan to market in the United States. B&S has 44 plants in Europe using 55 incinerator units. Forty-three of the units employ energy recovery. B&S has been selected as the preferred contractor for a 150 TPD plant in Fayetteville, Arkansas.

Katy Industries, Inc., the parent company of Fulton Iron Works, has had a marketing agreement with Seghers Engineering of Brussels, Belgium since 1979. Seghers has 15 plants in Europe wherein 16 out of 33 furnaces are used for waste-to-energy production. Katy-Seghers is the selected vendor for a 400 TPD system in Davis County, Utah; a 400 TPD system in Warren/ Washington County, New York; and a 480 TPD system in Savannah, Georgia.

Waste Management had been the general marketing agent for Volund in the United States since 1978. The Volund technology has been active in this country since 1943, when it was used in the first of 13 incinerators sold here. Two of these plants included energy recovery. Volund has 105 plants and 215 furnaces in operation. Of these, 94 are energy recovery units. Tampa, Florida selected Waste Management, with Volund technology, as the contractor for its waste-to-electricity system. In early 1984, Volund USA (VUSA), which was formed in 1978 as a technology transfer company for the System Volund waste-to-energy technology, separated its own marketing group from Waste Management. Volund USA now markets in the United

**Table 5–5. Refractory-Lined
Incinerator Vendors in the
United States.**

VENDOR	PROCESS
Lahontan Sacramento, CA	Bruun & Sorensen (Denmark)
Katy-Seghers Div. Fulton Iton Works St. Louis, MO	Seghers (Belgium)
Volund USA Ltd. Oak Brook, IL	Volund (Denmark)
Morse Boulger, Inc. New York, NY	"Kascade" Stoker (U.S.)

States, with the exception of a few geographical areas which are licensed by Waste Management.

Morse Boulger, Inc. (M-B) was established in 1890 as a partnership between Messrs. Morse and Boulger. Between 1932 and 1968, M-B constructed 33 municipal incinerators. In 1952, M-B installed the first modern cogeneration municipal refuse plant (700 TPD) in the United States in the Town of Hempstead, New York. In the late 1960's, M-B developed the "Kascade" stoker, and it has been installed at nine municipal sites over the past 15 years, including the waste-to-energy plants in Glen Cove, New York (250 TPD) and Harrisonburg, Virginia (100 TPD). In 1983, M-B became a subsidiary of Montenay International Corporation of New York, a firm which provides operation and maintenance services in the waste-to-energy field.

Modular Incineration

Controlled air incineration of solid waste using two or three combustion stages (i.e., modules) is commonly referred to as modular incineration. Capacities for single units range from 20 up to 160 TPD of MSW.

As depicted in Figure 5–5, modular units consist of a primary and a secondary (afterburner) combustion chamber connected to a downstream boiler system. The combustion of the waste fuel is accomplished by controlling the amount of air (i.e., oxygen) introduced into the chambers. One of the major benefits of this type of unit is the minimal field installation and erection work required since the units are usually shipped as large, preassembled components with wiring, plumbing, and instrumentation typically factory-assembled. However, increases in unit sizes (50 TPD capacity and larger) have led to shipments of large subsections which require more field erection but allow for major in-shop fabrication and assembly for better construction control.

Typically, hydraulically operated, ram-type feeders transfer the refuse into the combustion chamber. After entering the combustion chamber, waste is moved through the primary chamber by two general methods. Smaller systems (i.e., less than 10 TPD) use the force of the ram feeder pushing new waste into the primary chamber to move the waste in the primary chamber forward. A more common method employs the use of an internal ram-type stoker system or small reciprocating grate. The latter type of grate is similar to the reciprocating stoker assembly found in large field-erected direct combustion units. The majority of the vendors use internal transfer rams. The action of these rams, combined with the declining step arrangement of the furnace floor, helps to agitate the waste and expose new surfaces for combustion.

Combustion air for the grate system is provided by a forced draft (FD) fan

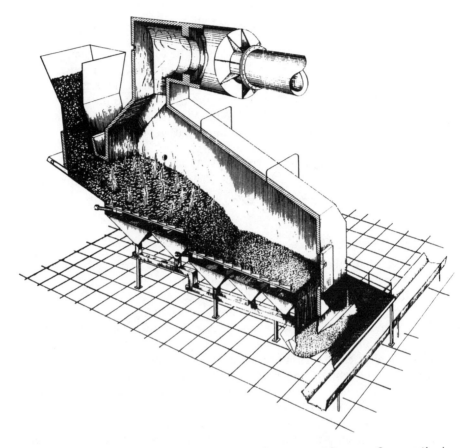

Figure 5-5. Modular incineration facility. *(Courtesy of Synergy Corporation)*

located upstream of the stoker section of the energy recovery unit. The FD fan directs air from underneath the grate system into the combustion chamber and forces the air to come in contact with the refuse. Combustion activity in the primary combustion chamber is sustained in either a substoichiometric state or an excess air condition, depending on vendor design.

Starved air systems provide for the combustion of any volatile gases produced by the substoichiometric air conditions in the excess air environment of the secondary or afterburner section. Because temperatures in the secondary chamber are in excess of 1,800 to 2,000°F, complete combustion of any volatile gas is ensured. This afterburner process acts as a crude air pollution control device.

Temperatures in the primary combustion chamber range from 1,500 to

1,800°F. The hot gas from combustion is used as the heating medium in a waste heat boiler located downstream of the secondary combustion chamber. This boiler is a smaller version of that installed on a refractory-lined mass burner. This unit may be fitted with a superheater section to produce high-quality steam. System thermal conversion efficiencies are typically in the range of 55 to 60 percent.

In many of the original modular installations, the secondary chamber afterburner was the only means of air pollution control. However, due to stricter federal and state emission standards, additional air pollution control devices are being installed on the discharge stream of most new units. Small prefabricated systems firing with excess air require an air pollution control device such as an electrostatic precipitator (ESP) or baghouse due to the increase in the quantity of air. The residue from combustion is discharged into a water quench trough for ash removal by a drag chain conveyor situated at the bottom of the trough.

General characteristics of modular incineration systems include:

- Economic viability in small waste generation areas
- Shorter construction time of 12–24 months, compared to larger field-erected systems with up to 24–36 month construction period
- Redundancy through the use of several smaller-sized units for handling larger plant capacities
- Flexibility in addressing various potential energy markets with system sizing

Background. While smoke from incinerators may have been tolerated around municipal incinerators in the past, there was less acceptance in other applications such as apartment buildings, hospitals, and commercial centers. To remedy the smoke problem, the small controlled air incinerator, now referred to as a modular incinerator, was developed in the late 1950s. However, the more clean burning, but more expensive, controlled air incinerators failed to have an impact on waste disposal practices until the promulgation of air quality regulations in the late 1960s and the passage of the Clean Air Act.

Like conventional field-erected refractory units, early refractory-lined modular incinerators were designed to simply reduce waste volume while also reducing particulate emissions. The first energy recovery application was in 1973 at St. Joseph's Hospital in Hot Springs, Arkansas. Discards from the hospital were used to fuel the system.

The first modular incineration application that included energy recovery using MSW was in Siloam Springs, Arkansas in 1975. Steam for a local food processing plant was produced by this facility. The small size of the operation (21 TPD) and the seasonality of the steam demand eventually led to the termination of steam production and abandonment of the facility. In 1977, the first continuous operation modular incinerator began operations in North Little Rock, Arkansas; see Table 5-6. This 100 TPD plant produces steam for a neighboring company—Koppers Industries' Forest Products Division.

Between 1975 and 1980, 11 modular style MSW-fired systems were built for energy recovery with steam sales to local industries. None of these facilities produced electricity. However, electricity was designed to be cogenerated by the energy markets from the steam purchased from two modular systems. These two systems are in Windham, Connecticut and Collegeville, Minnesota. The Windham energy market recently (1983) closed, and the City, which operates the energy recovery plant, is installing electrical generation equipment. Three other modular plants became operational in 1981 and 1982. None of these additional modular systems involved power generation or cogeneration, but they were larger in size (i.e., 200 and 240 TPD capacities). In 1983, a 112.5 TPD modular unit became operative in Cattaraugus County, New York and provides steam to a nearby cheese processing market. Facilities that also came on line in 1983 were located in Waxahachie, Texas; Red Wing, Minnesota; Park County, Montana, and several other locations. In 1983, a 64 TPD industrial waste–fired modular unit located at Zeeland, Michigan (Herman Miller, Inc.) started producing steam that is run through a small back-pressure turbine generator set and then produces process and heating steam for the industrial complex. Also, in early 1984, a large 300 TPD, four unit modular facility providing steam to an energy market came on line in Tuscaloosa, Alabama.

Current Status.   At the present time, numerous modular incinerators are under construction or contract, or vendors have been selected for negotiation. Four of these plants—Alameda, California; Rutland, Vermont; Springfield, Massachusetts; and Portland, Maine—will generate electricity as their output. A new 200 TPD plant in Oneida County, New York will cogenerate steam and electricity as does the 200 TPD plant in Oswego County, New York. Studies have been conducted to implement cogeneration as retrofit projects at the 200 TPD Auburn, Maine plant and the 108 TPD Durham, New Hampshire plant.

Future Developments.   A future development of particular interest from a power generation perspective is that modular incinerator system vendors are offering and guaranteeing systems that are capable of producing higher-quality steam conditions than in the past. These newer systems are claimed to be suitable for reliable power generation.

Table 5–6. Waste-to-Energy Systems: Modular Incinerators.[a]

YEAR OF START-UP	LOCATION	DESIGN CAPACITY (TPD)	ENERGY MARKET
1975	Siloam Springs, AR	20	—
1975	Blytheville, AR	50	—
1977	North Little Rock, AR	100	Koppers Co., Inc.
1979	Salem, VA	100	Mohawk Rubber Co.
1980	Crossville, TN	60	Crossville Rubber Products
1980	Dyersburg, TN	100	Colonial Rubber Works
1980	Durham, NH	108	University of New Hampshire
1980	Genesee Township, MI	100	(Purchased by General Motors)
1980	Lewisburg, TN	60	Heil-Quaker Co.
1980	Osceola, AR	50	Crompton Osceola, Inc.
1981	Auburn, ME	200	Pioneer Plastics Co., Division of Libbey-Owens Ford, Inc.
1981	Cassia Co., ID	50	Simplot Co.
1981	Johnsonville, SC	50	Wellman Industries
1981	Batesville, AR	50	General Tire and Rubber Co.
1981	Pittsfield, MA	240	Crane & Co., Inc.
1981	Newport News, VA	40	Ft. Eustis
1981[b]	Collegeville, MN	64	St. John's University
1982[b]	Windham, CT	108	Kendall Co., Inc.
1982	Waxahachie, TX	50	International Aluminum Extruders
1982	Portsmouth, NH	200	Pease Air Force Base
1982	Miami, OK	108	B.F. Goodrich
1982	Red Wing, MN	72	S.B. Foot Tanning Co.
1982	Park Co., MT	72	Burlington Northern Railroad
1983	Cuba (Cattaraugus Co.), NY	112	Cuba Cheese Co.
1983	Huntsville, AL	50	U.S. Army, Redstone Arsenal
1984	Pascagoula, MS	150	Morton-Thiokal Co.
1984	Tuscaloosa, AL	300	B.F. Goodrich
1984[b]	Oswego Co., NY	200	Armstrong Cork
1984[b]	Oneida Co., NY	200	Griffis Air Force Base
1984	Ft. Dix, NJ	80	Ft. Dix

[a] Source: Gershman, Brickner & Bratton, Inc.
[b] Electric generation or cogeneration.

Vendors. The modular incinerator companies have been primarily U.S. companies but, recently, several French firms have made inroads into the marketplace. Those vendors who are active in providing systems that process MSW into salable energy are listed in Table 5–7. There are numerous other modular incinerator vendors who do not actively compete in the municipal solid waste processing marketplace.

Table 5–7. Modular Incinerator Vendors
in the United States.

VENDOR	LOCATION
Basic Environmental Engineering, Inc.	Glen Ellyn, Illinois
Cadoux (French process)	Memphis, Tennessee
Comtro, division of John Zink	Tulsa, Oklahoma
Consumat Systems, Inc.	Richmond, Virginia
Synergy, Inc.	New York, New York
Clear Air Inc. (R. W. Taylor)	Ogden, Utah
Sigoure Freres (French process)	Washington, D.C.
Vicon Recovery Systems, Inc.	Butler, New Jersey

Basic Environmental Engineering, Inc. is a privately owned company with its headquarters in Glen Ellyn, Illinois, a suburb of Chicago. Founded in 1965 as Basic Engineering, the company at that time focused on the engineering, design, and sales of industrial waste incinerators. The first Basic incinerator was built in 1970, and additional industrial waste-to-energy units have been built since then. Two Basic systems for use with MSW are in operation: Collegeville, Minnesota (64 TPD) and Prudhoe Bay, Alaska (110 TPD). Basic has also sold several energy recovery units to industrial/commercial/hospital users over the past five years.

The incinerator sizes offered by the company range from 24 to 150 TPD on a 24 hour per day basis. All Basic units are shop fabricated with only limited field construction required. The Basic system has both a waterwall and a waste heat boiler recovery mechanism, which differs from other systems of this type that only offer the latter energy recovery approach.

Cadoux, S.A. is a large and diversified French manufacturing firm. Early in 1978, Cadoux acquired Athanor as part of a continual expansion and diversification program. Athanor had developed and built many of the municipal waste incinerators for the smaller communities in France, some dating back to 1971.

In June 1981, Cadoux, Inc. was incorporated under the laws of the State of Tennessee. It is a wholly owned subsidiary of Cadoux, S.A. Cadoux, Inc. is engaged in the planning, designing, financing, and construction of small-scale waste-to-energy projects. Cadoux notes that it has 13 waste-to-energy plants operating or under construction, with small-scale U.S. projects in Delaware County, Pennsylvania (50 TPD) and Cleburne, Texas (100 TPD).

Comtro is a division of John Zink of Tulsa, Oklahoma and was acquired from Sunbeam in 1984. Founded in 1958, Comtro was purchased by Sunbeam in 1973. In late 1981, Allegheny International purchased the Sunbeam Corporation. The company designs and manufactures package incinerators. The Comtro process is marketed on a turnkey basis. The first incinerator for industrial waste was installed in 1968. In 1975, the first unit with energy recovery, also for use with industrial discards, was built. A Comtro energy recovery system is operating at the Jacksonville Naval Air Station (75 TPD) that uses the refuse from the base. Comtro has over 20 waste-to-energy units installed, mainly to industrial clients. The waste-to-energy incinerator sizes offered by the company range from 3.5 to 50 TPD on a 24 hour per day basis. All Comtro units are shop fabricated with only limited field construction required.

Consumat Systems, Inc., a publicly owned Virginia corporation with over 15 municipal waste-to-energy plants on-line, raised considerable amounts of cash by selling a significant stock interest in the firm in July 1983 to Melvyn L. Bell of Little Rock, Arkansas. Additionally, Consumat started teaming with larger firms such as RECO Industries, Inc. to provide more financial backing for many of the private deals and long-term operations commitments, as well as to cover the potential liabilities of the projects that Consumat wants to pursue in the marketplace. By way of technology advancements, their new 75 TPD modules were installed at the 300 TPD Tuscaloosa, Alabama facility which began operations in February 1984.

Synergy Systems Corporation is a wholly owned subsidiary of McMullen Holdings, Inc., a privately held corporation. Through its predecessor, Synergy has been involved in small-scale energy recovery since 1974. Synergy–Clear Air, from August 1979 to January 1982, was involved in marketing and designing three operating waste-to-energy plants (Waxahachie, Texas; Cattaraugus County, New York; and Dade County, Florida International Airport) ranging from 60 to 112 TPD.

R.W. Taylor Group/Clear Air, Inc.: The R.W. Taylor Group consists of the R.W. Taylor Investment Group, the R.W. Taylor Steel Company, and the R.W. Taylor Construction Company. Clear Air, Inc. is a publicly owned corporation with the majority of its outstanding stock owned by the Taylor Group.

In 1966, the R.W. Taylor Steel Company entered into a turnkey-type contract with Clear Air Reduction Corporation of Chicago, Illinois for the design, fabrication, and construction of a 300 TPD incinerator for Weber County in Ogden, Utah. R.W. Taylor became very interested in incineration and eventually bought the Clear Air patents. At present, R.W. Taylor has a 112.5 TPD energy recovery/steam-generating unit operating in Cattaraugus County, New York and is in final construction on a 200 TPD energy recovery cogeneration facility in Oneida County, New York.

Sigoure Freres, S. A. of France (SF) produces a line of turnkey plants using modular incinerators up to a maximum of 132 TPD. They are reported to be the leading manufacturer of small-scale municipal incinerators in France.

The Sigoure Group is controlled by SIGFI, S.A. (Sigoure France). The Sigoure brothers own 90 percent of SIGFI, with the other 10 percent held by an agency of the French government, IDI, which takes equity positions in fast-growing French firms to encourage their continued expansion. The technology available consists of either a step grate design (four models from 15 to 45 TPD capacity) or a rotating grate design (five models available from 30 to 100 TPD design capacity).

The marketing of the Sigoure Group in the United States is through Sigoure U.S. Associates of Washington, D.C. SF was selected for its first U.S. facility in September 1982. It is a 150 TPD, two line rotary hearth system that started up in October 1984. In July 1983, SF was low bidder on a 25 TPD solid waste and sludge-to-energy project in Sitka, Alaska. SF was selected as part of a full-service contractor team in November 1984 for the 225 TPD cogeneration project in St. Lawrence County, New York.

Vicon Construction Company is a privately owned company with its main office in Butler, New Jersey. The company was founded in 1966 as a construction company. In 1978, Vicon entered into an exclusive license agreement with Enercon Systems, Inc., an incinerator manufacturer. Vicon Recovery Associates is a limited partnership comprised of the Vicon Construction Company and eight limited partners. This partnership was formed to act as the contractor for their 240 TPD Pittsfield, Massachusetts project. Vicon has been selected as an equipment subcontractor to Crouse for a four unit, 600 TPD plant in Wilmington, Delaware which has just completed financing.

Problems

Energy recovery via direct incineration is not without problems. Refuse contains many constituents that should not be burned. Glass, grit and silt, and metallic objects are very abrasive materials which cause maintenance problems with the stoker. The firing of plastics and other halogenated hydrocarbons provides for very high heating values but releases corrosive compounds in the form of hydrogen chloride (HCl), hydrogen fluoride (HF), and chlorine (Cl_2). These chemicals, through a complex chemical reaction, attack boiler tubes and, over time, reduce the wall thickness of a tube until it eventually fails. It should be noted that the only metal that is completely resistant to all of these chemicals is titanium. For economic reasons, one cannot construct a boiler with titanium tubes. Reasonable success has been achieved using an impregnable refractory lining made of silicon carbide (SiC). This material is impervious to chemical attack, but should the SiC temperature exceed 2,000°F, the SiC will tend to become soft, which causes ash to deposit on its surface. Operating personnel must monitor clinker growth on SiC;

otherwise, unanticipated downtime due to excessive slagging will occur. Another more expensive method of protecting boiler tubes from corrosion is the use of composite tubing. This type of tubing is extruded specifically such that stainless steel is on the fire side of the tube, while carbon steel is the water-side material.

Flame patterns inside the furnace should be controlled to avoid direct flame impingement on boiler tubes. Flame impingement will cause thermal corrosion of the boiler tubes, while creating a reducing atmosphere at the tube-flame interface. This reducing atmosphere will cause corrosive materials (HCl, HF, Cl_2) to attack the boiler tubes.

Because of the large amounts of fly ash generated during the combustion of refuse, the superheaters and economizers must be cleaned by soot blowing more often than for coal-fired boilers. Excessive use of the soot blowers, particularly those that use steam, may cause erosion on the tubes, eventually leading to tube failures. In addition, the fly ash is very corrosive as well as erosive. This ash contributes to the destruction of superheater and economizer tubes. These problems can be partially alleviated by installing stainless steel shields on the tubes.

Because unprocessed refuse is an abrasive material, care should be taken so as to select the proper grate material/alloy. The proper alloy should provide for mechanical, as well as thermal, durability. Failure in doing so will only result in unwanted furnace downtime.

Operating personnel must try to insure that refuse is completely burned at the end of the grate cycle, otherwise smoldering, and in some cases, burning refuse will enter the ash disposal system. If this situation persists, downtime resulting from the maintenance of the disposal system will occur. Operating personnel must also take care to remove all large nonburnable and non-processible objects (i.e., white goods, "bulkies") before introducing refuse into the feed chute system of the furnace. Failure to do so will result in damage to the integral parts of the waste incineration system.

REFUSE-DERIVED FUEL (RDF)

RDF Processing

Dry and wet systems are the two approaches available to process solid waste to produce an RDF.

Dry RDF Processing. "Fluff," "densified," and "dust" are terms used to describe the type of products that can be produced with dry RDF processing. The production process for all three is similar through the fluff stage. Additional processing is required to obtain the densified and dust types of

RDF. Market requirements typically determine the type of RDF that is produced.

Fluff RDF. Fluff RDF has a relatively small particle size. Depending on the application, the typical size may vary from about 4 inches for use in a spreader stoker boiler down to approximately 0.75 inch for suspension-fired boilers. Fluff RDF has been used in both dedicated boiler and co-fire applications.

Although several configurations are in use, a typical process system to produce fluff RDF is shown in Figure 5-6. Waste is unloaded into a pit or onto the receiving floor and then conveyed to a flail mill. The flail mill tends to separate the waste mass for further downstream processing as well as open any bags containing refuse. Magnetic separation of the ferrous metal is the next stage in the processing. An additional RDF processing step is conducted by the use of a trommel screen. This equipment is a rotating, slightly inclined screen that allows waste to flow through the cylinder in a lift-and-drop fashion while removing many of the remaining small-sized noncombustible materials through the screen holes.

The waste that passes through the trommel is fed into a shredder, which reduces the particle size of the waste. The shredder waste is fed into an air classifier. An air classifier is typically a vertical column in which an upward flowing air current separates the lighter, more combustible materials from the remaining noncombustibles. During this separation process, some of the combustible material is diverted from the RDF and exits the system with the inorganics. This lowers the total energy available from a given quantity of MSW. The loss, however, tends to be minor. If the smaller (0.75 inch) particle size is desired, then the waste would be shredded again at this point.

The lighter material obtained from the air classifier is fluff RDF. A problem common to all types of fluff RDF is limited storage capacity. The material will compress under its own weight, which makes retrieval for use difficult. In general, fluff RDF storage is limited to 24 hours.

The configuration shown in Figure 5-6 has evolved from other processing

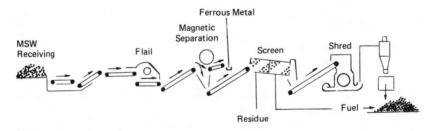

Figure 5-6. RDF feed system

systems. The common first generation systems involved a shredder followed by an air classifer. The heavier, noncombustible materials could be further processed to recover ferrous or other materials. A problem with shredding as the initial process is that inert material, such as glass, tends to become imbedded or impregnated in the combustible material. Subsequent separation processes fail to dislodge the material; the result is an erosive quality RDF with a high ash content, which can cause short-lived pneumatic conveyor tubes and boiler slagging problems. This is particularly a problem in co-firing with coal in utility boilers. Such units operate at higher combustion temperatures than those normally encountered in a solid waste incinerator, which causes the slagging to occur.

Densified RDF.   Densified RDF (d-RDF) is produced by compressing fluff RDF into pellets or briquettes. This requires that fluff RDF be produced in the smaller (0.75 inch) particle size. Densified RDF can be co-fired with coal in lump form and is the usual reason for producing this type of RDF. Longer-term storage and transportation advantages are important features of d-RDF.

Dust RDF.   Fluff RDF can be treated with an embrittling agent (sulfuric or hydrochloric acid) and then fed into a ball mill (hot steel balls shatter the embrittled waste in a rotating chamber) to produce dust RDF. This type of RDF can be used in a suspension-fired boiler as well as mixed with oil for use in a conventional oil-fired boiler. In view of previous processing problems with this form of RDF, little interest and no commercial development funds are being given to its application at present.

Wet RDF Processing.   A wet pulp process is used to make wet RDF. Water is fed into a pulper, which resembles a large blender that mixes water and solid waste. The combustible materials form a slurry which exits from the bottom of the pulper. Nonpulpable materials (glass, metal) are rejected from the pulper. The slurry is dried to 50 percent solids before being used as a boiler fuel. An advantage claimed by this process is the elimination of explosions and dust inherent in dry processing. However, odor problems have resulted from the use of recycled water in the pulping process.

Refuse-Derived Fuel Co-fired With Coal

Refuse-derived fuel fired in conjunction with pulverized coal in a suspension-type boiler is known as "co-firing" RDF with coal. Unlike the mass burning or RDF dedicated boiler concepts where only refuse is fired in the furnace, the combustion of refuse in the co-fire application takes place above the grate in a fireball created by the combustion of pulverized coal. If the proper machinery and combustion conditions exist, RDF can supply up to 20 percent or more of fuel heat value in existing and appropriately modified systems. Figure 5–7 depicts a typical suspension-fired coal boiler.

In order to fire RDF at the chosen boiler site, the following systems would be required:

- Receiving of MSW and a processing facility for RDF
- Transportation of RDF
- RDF receiving facilities
- RDF storage facilities
- Firing facilities

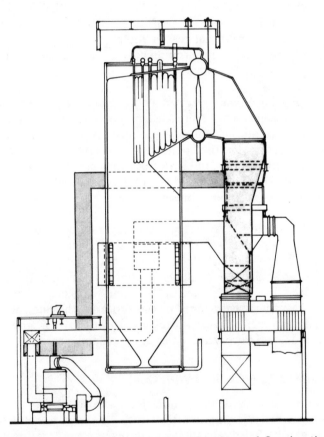

Figure 5–7. Suspension-fired coal boiler. *(Courtesy of Combustion Engineering, Inc.)*

In order to co-fire RDF, it is usually transferred pneumatically to an interim surge bin where it is conveyed pneumatically to the firing system. The use of pneumatic rather than belt-type conveyors is warranted in this type of application due to the light, fluffy nature of the fuel. Belt-type conveyors are expensive and do not allow for a complete environmental seal. The use of pneumatic conveyors allows for an environmental seal, as well as transporting RDF efficiently.

The surge bin acts as a small fuel reservoir for minor mechanical breakdowns involving upstream equipment (i.e., crane or front-end loader). The surge bin is equipped with either rotating screw-type feeders or drag conveyors which would transfer the RDF into a pneumatic feeder. The pneumatic feeder, which consists of a transfer tube, fuel hopper, and high-pressure blower, transfers RDF to a surge bin/holding silo. The silo is constructed to have sufficient volume to ensure that a constant supply of RDF can flow to the boiler should either the surge bin or the pneumatic conveyor/feeder fail for a small period of time. A schematic of the feed system is given in Figure 5–8. The silo may be equipped with either a multiple rotating screw or a drag conveyor arrangement to ensure proper discharge from the silo base.

The conveying system discharges RDF into transfer tubes which in turn discharge into the metering fuel feeders. The metering feeders are equipped with high-power fans that blow refuse into the combustion chamber of the boiler. The discharge port from which RDF is injected into the combustion chamber is equipped with a variable pitch nozzle. This feature enables operators to direct fuel at the proper spot in the combustion chamber to facilitate complete and efficient combustion of RDF. The nozzles are located above the pulverized coal feeders. This arrangement allows the RDF to be incinerated in the coal fireball. Feeding RDF into the combustion chamber

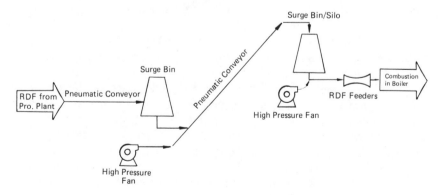

Figure 5–8. RDF transport system.

below the coal feeders will result in poor fuel combustion, leading to boiler control problems.

Due to the heterogeneous nature of RDF, some particles may not completely burn in suspension and would then fall into the boiler ash hopper. Failure to control this phenomenon may lead to ash bridging over the ash pit and unanticipated clinker growth in the pit. The installation of dump grates in a suspension-type boiler's ash pit usually will eliminate this problem by allowing the unburned refuse to burn completely on the grate. Ash from the combustion of coal and RDF either falls directly into the ash pit (approximately 20 percent of the total ash generated) or is carried in the gas stream as fly ash to be removed eventually by the downstream air pollution control equipment.

Depending on the nature and quantity of RDF fired in the boiler, air requirements for combustion have to be closely examined. The addition of air into the boiler from the RDF feeders, along with increased combustion air requirements, may warrant increases in both forced draft and induced draft fan capacities. In addition, a greater quantity of ash will be generated due to RDF combustion. This may require an upgrade of the existing ash handling system. Efficiencies of pulverized coal–firing steam-generating units range from 80 to 90 percent. Because the RDF fuel, if combusted properly in a converted pulverized coal–firing boiler, burns much in the same manner as pulverized coal (i.e., in suspension), boiler efficiencies remain generally unaffected or may drop only 1 to 2 percent.

Refuse-Derived Fuel Dedicated Boiler

The use of the term "dedicated boiler" in the waste-to-energy field typically refers to a spreader stoker unit (see Figure 5-9). Such units are designed to accept RDF as the primary fuel. Like the mechanical grate direct combustion systems, an RDF dedicated boiler will have water tubes lining the combustion chamber for heat recovery.

Specific design benefits associated with the firing of RDF in dedicated boilers include better fuel predictability than the MSW and specially designed furnace areas, fans, and tube spacings. Also, air pollution and ash handling equipment can, and should, be designed specifically for the use of RDF as a primary fuel. Capital cost per ton of RDF combusted is generally higher for dedicated boiler systems than with co-fire systems. This is because the cost of a new boiler is usually more than that incurred for retrofitting an existing unit. Despite the greater degree of processing involved, total capital costs typically are competitive with mass burning waterwall incineration, primarily because of reduced combustion air and, consequently, reduced boiler volume and air pollution control (APC) requirements. Because of the

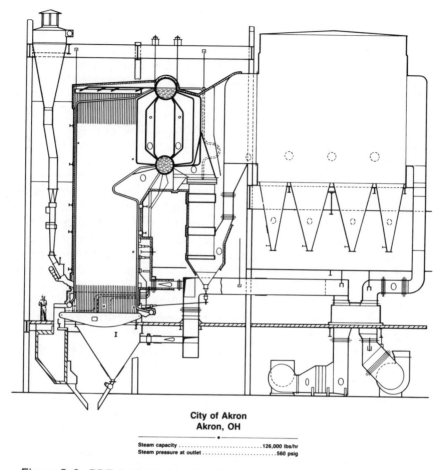

City of Akron
Akron, OH

Steam capacity .... 126,000 lbs/hr
Steam pressure at outlet .... 560 psig

Figure 5-9. RDF dedicated boiler. *(Courtesy of Babcock & Wilcox, Inc.)*

substantially greater surface area of RDF, the more homogeneous characteristics of the feed, and the removal of some problem noncombustibles, combustion control and system efficiencies could be superior to mass burning
waterwall systems.

Operation of a dedicated boiler is very similar to that of a mass burning
waterwall unit, with the major differences located in the fuel feed system.
The dedicated boiler will receive fuel to its combustion chamber either by
gravity or by pneumatic feed. As refuse is fed into the boiler, the very light
dust particles are burned almost instantaneously in suspension. Then the ash
is carried in the gas stream and removed by air pollution control equipment.
Heavier particles fall to the front of the stoker. There they are combusted

quickly while the heaviest particles travel through the combustion zone to the rear of the boiler. They can remain on the grates for a longer period and burn completely; see Figure 5-10.

The stoker assembly is a moving grate unit made up of a number of grate bars. The speed of the stoker is adjustable to facilitate complete combustion on the grate. Some boilers are fitted with a coking plate at the rear of the boiler that starves the fuel of oxygen, in essence, creating a pyrolysis zone at the rear of the boiler. This plate also helps keep the flame away from the rear wall of the boiler, thus reducing the probability of tube wear and corrosion at that spot.

On the grate, hot air, supplied by the forced draft fan from an air preheater, initiates the combustion of refuse. As the fuel moves closer to the front end of the boiler, it is slowly combusted. Once it reaches the front of the boiler, depending on steam load conditions, an ash bed depth of 4 to 6 inches can be expected. The sandy, dry ash is discharged into an ash pit. The distance to which fuel is distributed into the boiler is controlled by the refuse fuel spouts. By controlling the amount of air to the spouts, RDF can be effectively moved forward or backward in the boiler.

The combustion chamber of a dedicated boiler is normally very tall (50-70 feet) in order to facilitate the combustion residence time needed for suspension burning. Combustion air for the suspension burning is supplied by the overfire air system. This system of air nozzles is located between 2 and 15

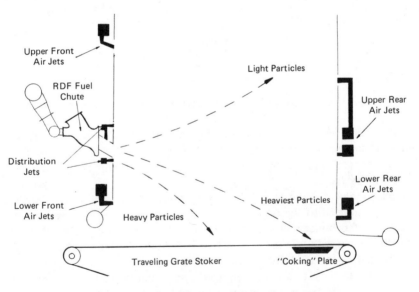

Figure 5-10. RDF feed distribution system.

feet above the stoker. It provides an adequate supply of air for the complete combustion of suspended particles, as well as keeping the flame away from the lower waterwall sections of the furnace where it is most intense. This air is delivered from the air preheater to the overfire air fan (OFA fan) which distributes the air to the overfire nozzles. Overfire air is supplied at a positive pressure of approximately 20 to 30 inches of water, compared to a positive pressure of 1 to 2 inches of water under the grate. This high-pressure air is needed to provide air to the flame in the center of the boiler and to ensure complete suspension burning.

The hot gases of combustion are carried away by the induced draft fan which pulls the gas from the combustion chamber through the boiler (convection tube bank), where it surrenders its heat to the boiler tubes and boiler water, and through the economizer to the air pollution control equipment.

As with mass burning units, boiler pressure should be kept at subatmospheric conditions to ensure that noxious fumes are retained in the boiler. Air supplied to the combustion chamber is in excess of 40–60 percent above stoichiometric conditions. As with waterwall incinerators, the circulating water serves to keep the combustion chamber cool, as well as to absorb some of the heat of combustion. As the hot gases rise through the boiler furnace area, they pass through the superheater tube bank and heat the saturated steam to the desired degree of superheat. (As with waterwall units, the degree of superheat is controlled by attemperators on the superheater discharge header.) The hot gases then pass through the convection tube bank of the boiler, the economizer, and finally, the air pollution control system.

Steam, either superheated or saturated, is piped to a turbine-generator set or to ultimate steam users via a common distribution header. Steam generated from such units ranges from saturated steam to superheat conditions of 750°F/600 psig. A few exceptions to this exist, one being Occidental Energy Company's RDF cogeneration system at Niagara Falls, New York, which operates at 750°F/1,250 psig steam.

Construction of a dedicated boiler is very similar to that of a waterwall unit. Also, water treatment in a dedicated boiler is very similar to that of a waterwall boiler.

Ash that is discharged into the ash pit may be removed by a wet quench or dry system. The use of a dry system, as previously explained, can be a health hazard. In addition, due to the large amounts of air entering the furnace when ash is pulled, the delicate balance of air being supplied for combustion is upset. This, in turn, upsets the combustion process. The use of a wet quench system will allow for a positive seal between the atmosphere and the boiler, thus eliminating the possible occurrence of a hazardous condition and keeping the boiler in operation under subatmospheric conditions. Economic considerations normally dictate whether a wet or a dry system is used. Generally, because of the ancillary equipment associated with this process (conveyors, quench tank, etc.), wet systems are more expensive than dry systems.

As with any process, the control and operation of dedicated boilers must be finetuned. Depending on furnace and combustion control, energy recovery efficiencies may be in excess of 70 percent. Fuel feed rates lower than 50 percent of design capacity will result in poor combustion, as well as operating problems not experienced at higher loads.

The features of an RDF dedicated boiler system allow for the following:

- Combustion of a homogeneous fuel
- Minimal need for supplementary (fossil) fuel
- Boilers and supplementary equipment specifically designed for the combustion of refuse
- Boiler efficiencies in the range of 65 to 75 percent

Background

Dry RDF Processing/Combustion.   At about the same time that water-wall combustion was being applied to solid waste in Europe, interest was also developing there on the use of shredders to reduce the volume requirements of MSW in landfills. The first shredding system in the United States began operation in 1966 in Madison, Wisconsin and was based on European experiences. This system was developed to demonstrate the viability of disposing of shredded waste without daily cover. Several other shredding systems began during this period. Recovery with these systems was limited to ferrous metal that could be easily recovered through magnets. These initial efforts at shredding prior to landfill eventually led to the development of RDF processing systems both in the United States and in Europe. Madison, for example, chose to build an RDF system when its landfill began to reach capacity in the mid-1970s.

The first demonstration using shredded MSW as a fuel began in 1972 in St. Louis, Missouri. The fluff RDF was co-fired with pulverized coal in two suspension-fired boilers owned by the local utility—Union Electric Company. As might be expected with any initial demonstration project, numerous problems were encountered. The primary problem in terms of electrical production was that the RDF caused a significant increase in ash quantities. Modifications of the RDF production process, such as air classification, were made to reduce the content of glass and other inert materials in the RDF.

Originally funded as a 300 TPD demonstration plant and proof-of-concept system under a short-term grant from the U.S. Environmental Protection Agency (EPA) in 1970, the St. Louis project was discontinued in 1976. An attempt to make the project a 4,000 TPD full-scale system was based on the local utility's passing its costs onto the rate payers. Voters rejected this option in a referendum. There are several RDF plants in Europe producing a fuel for co-firing in cement kilns.

The first commercial-scale RDF plant in the United States began operations in March 1975 and was built in Ames, Iowa. A fluff RDF is produced in Ames and co-fired with coal in existing boilers operated by the City's utility company. Since the Ames plant became operational, 20 RDF plants using the dry processing approach have been built or are under construction in the United States; see Table 5-8. The record of these plants in producing a suitable energy product for the utility customer is poor. Eight of the plants are currently closed. The plant that was producing a dust RDF was closed because the company then operating the facility filed for bankruptcy. In addition to Ames, a small RDF co-firing plant in Madison, Wisconsin and a new facility in Lakeland, Florida are operating as designed.

Table 5-8. Waste-to-Energy Systems: RDF Systems.

YEAR OF START-UP	LOCATION	DESIGN CAPACITY (TPD)	ENERGY MARKET
1975[a]	Ames, IA	200	Ames Municipal Electric System (electricity)
1976[a]	Baltimore Co., MD	1,200	Baltimore Gas & Electric Co. (electricity)
1977	Chicago, IL	1,000	Commonwealth Edison (terminated)
1977	E. Bridgewater, MA	300	—
1977	Milwaukee, WI	1,600	Wisconsin Electric Power Co. (terminated)
1978	Hempstead, NY	2,000	Long Island Lighting Co.
1978	Lane Co., OR	500	(Firm market never developed)
1979	Akron, OH	1,000	District steam loop (cogeneration being considered)
1979	Bridgeport, CT	1,800	—
1979	Duluth, MN	400	Codisposal (presently firing wood chips)
1979[a]	Madison, WI	400	Madison Gas & Electric Co. (electricity)
1979	Tacoma, WA	500	(Firm market never developed)
1980[a]	Monroe Co., NY	2,000	Rochester Gas & Electric Co. (electricity)
1981	Albany, NY	750	District steam loop

(cont.)

Table 5–8. (continued)

YEAR OF START-UP	LOCATION	DESIGN CAPACITY (TPD)	ENERGY MARKET
1981[a]	Niagara Falls, NY	2,200	Hooker Chemicals & Plastics Corporation (cogeneration)
1982[a]	Dade Co., FL	3,000	Florida Power & Light Co. (electricity)
1982	Henrico Co., VA	200	—
1983[a]	Wilmington, DE	1,000	Delaware Power and Light Co. (electricity)
1983[a]	Lakeland, FL	300	Lakeland Power Co. (electricity)
1983[a]	Columbus, OH	2,000	Columbus Municipal Power System (electricity)
1985[a]	Haverhill and Lawrence, MA	1,300	Industrial park complex (cogeneration)
UNDER CONSTRUCTION			
1987	Norfolk, VA	2,000	U.S. Naval Base (Norfolk)
1987[a]	Rochester, MA	1,500	Commonwealth Electric Co. (electricity)
1987[a]	Detroit, MI	3,000	District steam loop (cogeneration)
1987[a]	Honolulu, HI	1,500	Process steam (electricity)
1987[a]	Hartford, CT	1,500	Steam to utility turbine (CL & P)

[a] Electric generation or cogeneration.

As an alternative to co-firing RDF with coal in existing boilers, several communities have chosen to build boilers specifically designed for burning RDF alone. Included in this group are Akron, Ohio; Albany, New York; and Niagara Falls, New York (see Table 5–8). The original MSW dedicated boiler that has used this approach since 1969 is still in operation in Hamilton, Ontario. The Akron, Niagara Falls, and Hamilton projects all involve processing the waste through a shredder as the initial step. The Niagara Falls plant passes the steam through turbines prior to distributing the energy for process needs.

Wet RDF Processing/Combustion. In 1967, the landfill in Franklin, Ohio was beginning to reach capacity. Difficulty in siting a new landfill led to the involvement of Black Clawson and the development of a hydropulper system, widely used in the paper industry, for materials recovery from waste. The pulped paper and other organics were combined with sewage sludge and incinerated in a fluidized bed incinerator. Energy recovery was not part of this system. The federal EPA demonstration project was completed in 1976.

This process was originally conceived as a materials recovery process, including the production of a low-grade paper. However, the rise in energy prices in late-1973, due to the action of OPEC, caused Black Clawson to reconsider its objectives. Parsons & Whittemore, the parent company of Black Clawson, began marketing the system as an RDF processing unit that would be tied in with a dedicated boiler. A 2,000 TPD plant has been built in Hempstead, New York and a 3,000 TPD plant in Dade County (Miami), Florida. The Hempstead plant is infamous because dioxin, a carcinogen, was found in emissions from the boiler in 1980. The EPA eventually ruled that the amount of discharge was within an acceptable level. Although dioxin has been blamed for the closing of this plant, other factors related to the management of the plant—labor issues, odors, and operational problems—have had a significant role in its still being closed. Both plants were designed to produce electricity, and the Dade County facility is continuously operating and producing power for sale to the Florida Power and Light Company.

Current Status

Several RDF systems are being built or are currently under contract. Most recent generation RDF co-fire units have started up in Lakeland, Florida and Columbus, Ohio. Although these facilities are co-fired with coal, they are unique because the boiler systems are new. This approach avoids the problems associated with trying to retrofit an existing boiler (usually designed for a much different fuel specification) to burn a supplemental fuel. Both systems will be used for power production.

Another RDF development is the 1983 decision of Baltimore Gas and Electric Company to co-fire RDF with coal at its C.P. Crane power plant. The RDF is produced at the Baltimore County processing facility. The Baltimore County RDF processing facility operator had been attempting unsuccessfully to market the RDF for over five years.

An RDF dedicated boiler project owned by Refuse Fuels, Inc. started commercial operations in 1985 at Haverhill/Lawrence, Massachusetts. A 1,500 TPD processing plant will provide RDF to an existing power plant which added a new boiler to burn RDF. The system cogenerates electricity and supplies steam to an established industrial park complex. The industry will be carefully watching this project, as it is the latest generation of the RDF dedicated boiler concept.

The City of Detroit has selected Combustion Engineering (CE) to build an RDF dedicated boiler system. This facility will produce steam that will be fed into the downtown district heating steam loop. In addition, electricity will be cogenerated for sale to Detroit Edison. Other RDF dedicated boiler projects are planned in Honolulu, Hawaii; Saco/Biddeford, Maine; Palm Beach County, Florida; and San Francisco, California.

Future Developments

With RDF, future developments will probably be in the refinement of the processing phase so that a fuel with less ash and inerts is produced. This would most likely involve modifications to current process approaches rather than an entirely new method of RDF preparation. Another development probably will be the increased use of RDF in dedicated boiler applications. The trend is away from extensive materials recovery and RDF processing, and back to a simpler process train producing a larger RDF particle size.

Vendors

The leading system vendors in the RDF field are listed in Table 5–9.

Energy Answers is an independent firm located in Albany, New York. Its processed fuel/dedicated boiler technology is in operation in Hamilton, Ontario and Albany, New York. The Hamilton plant has been in operation since 1972. Additional ongoing activity of Energy Answers Corp. includes developing a 1,200 TPD project in Massachusetts known as the SEMASS project which was scheduled to be financed in late December 1984.

The Albany, New York project, which supplies RDF that is used to produce steam for a district steam loop, is operated by Aenco.

Parsons & Whittemore is the parent company of Black Clawson, which built the systems in Hempstead, New York and Dade County, Florida. These systems use the wet processing method. At present, the company has no other systems under construction or contract.

Table 5–9. RDF System Vendors in the United States.

VENDOR	HEADQUARTERS LOCATION
Energy Answers	Albany, New York
Aenco-Cargill	Albany, New York
Black Clawson (Parsons & Whittemore)	New York, New York
Combustion Engineering, Inc.	Windsor, Connecticut
Mason-Dixon Resource Developers	Henrico County, Virginia
Mustang RDF Co.	Oklahoma City, Oklahoma
National Ecology, Inc.	Los Angeles, California
(formerly Teledyne National)	
Kuhr Technologies	Hackensack, New Jersey
Tricil, Inc.	Akron, Ohio

Combustion Engineering (CE) has been promoting RDF processing systems that use the dry RDF approach. CE has been selected to build a 3,000 TPD RDF plant in Detroit, Michigan that would supply steam to a downtown steam loop. A 1,500 TPD CE plant in Hartford, Connecticut that would supply steam to an existing utility turbine facility has recently been financed; and CE has been selected for 1,500 TPD plants in Honolulu, Hawaii and San Francisco, California. The latter two plants have not yet been financed.

Developed under a grant from DOE, Mason-Dixon Resource Developers use a process that involves pressurizing MSW and then exploding it, which causes the waste to shred. A plant using this process was built in Henrico County, Virginia. This facility is presently shut down.

Mustang RDF Co. has been mentioned as the contractor for some projects. However, the company currently has no plants under construction or contract.

National Ecology, Inc., formerly Teledyne National, built the RDF processing system in Baltimore County, Maryland and operates it under contract to the County. In addition, Teledyne was the interim operator of the Akron, Ohio plant prior to that plant's being taken over and upgraded by a waste management service company—Tricil, Ltd.

Kuhr Technologies was organized in 1982 to develop waste-to-energy facilities. Kuhr is the developer for the Saco/Biddeford, Maine project financed in June 1985, which is to be designed, constructed and operated by General Electric. Kuhr has also organized Penobscot Energy Recovery Company, a Maine limited partnership, to develop a second waste-to-energy project in Bangor, Maine.

Tricil's initial involvement in the waste-to-energy business began in October 1977 when it was selected for the major upgrading and operations of the 600 TPD Hamilton, Ontario refuse processing and RDF dedicated boiler firing plant. The plant, which started up in 1972 and was operated by the Regional Municipality of Hamilton-Wentworth, was retrofitted and upgraded by Tricil in 1979, and it has been given a ten year operating contract.

In June 1981, Tricil was successful in obtaining a contract to upgrade and operate the 1,000 TPD City of Akron, Ohio refuse-derived fuel preparation plant and multiboiler energy facility. The Akron Recycle Energy System (RES) was commissioned in 1979 and experienced significant problems during the start-up and operations phases. Tricil Resources, Inc., a subsidiary of Tricil Limited, was selected as the contractor and completed extensive modifications and upgrading of the plant. It is now operating the facility.

As with the refractory systems, an alternative to the system vendor procurement is the selection of an architect/engineering (A/E) firm to design and build the plant. This approach has been used in several RDF projects.

Problems

As with direct incineration, energy recovery through the combustion of RDF has its problems. The preprocessing of MSW into RDF removes much of the unwanted, noncombustible materials. However, the remaining unwanted materials cause serious maintenance problems, particularly with the stoker. Aluminum, which is light enough to be carried in the refuse stream, will melt on the grate bars. The stoker grate bars require almost perfect alignment. This tramp aluminum finds its way to the grate bar/grate rail interface, causing wear and alignment problems. If this occurs, expensive shutdowns follow. Glass and silt, which may become impregnated in the refuse during preprocessing, will also cause accelerated grate wear problems.

First generation RDF processing plants experienced a host of problems, several of which still plague this technology today. These problems are:

- Explosions
- Dust in the processing plant
- Storage/retrieval problems
- Materials handling problems

Explosions have been a part of the refuse processing industry since its inception. In order to process a refuse stream of heterogeneous material, of different sizes and shapes, to one of uniform size, the material must be reduced in size. The method by which this size reduction is effectuated is through a high-speed/high-impact shredder. These units are fitted with hammers which essentially "push" refuse through grates (or screens) that determine the final product size. Because the incoming waste was not (and, in some cases, still is not today) preprocessed, many flammable liquid containers, and some compressed gas tanks, made their way to the high-impact shredding unit. As a result, explosions from the combination of flammable materials and high impacts occurred. Needless to say, an explosion of any magnitude will cause facility downtime, as well as being a serious threat to human lives. In order to reduce the probability of explosions occurring, many facilities have operators pick through the refuse to look for flammable, explosion causing containers. One facility has installed a low-speed shear shredder or rip shear to break up flammable containers prior to their entering the high-speed shredder. The Tricil RES facility in Akron, Ohio has installed shredder relief vents on its high-speed shredders so as to relieve the sonic pressure wave generated by an explosion. Nearly every facility operating high-speed shredders is equipped with the Fenwal Explosion Suppression System.

Dust created by the high-speed shredding of materials is a problem as well.

In order to alleviate this problem, the shredder building must be properly ventilated and not connected to any operator working areas.

The storage and retrieval of RDF is still a problem with facilities that choose to store the fuel over a lengthy time period. RDF tends to "compress" itself such that retrieval of the refuse at the bottom of a pile tends to be difficult. Storing RDF for lengthy periods of time causes the fuel to ferment slowly and to spontaneously combust.

Control of combustion is very important with dedicated units. Because a delicate balance of air is required, any excess tramp air will cause problems. Tramp air may be introduced through the ash pit, the refuse chute, or even through observation doors in the boiler. Because the induced draft fan is sized for a certain amount of air, any additional air taxes the capacity of the unit. Also, because the induced draft fan is not selective as to what it removes from the combustion chamber, it will most likely remove a mix of tramp air and hot furnace gas. This will lead to poor furnace pressure control. The leakage of tramp air will also cause poor RDF distribution in the firebox, thus creating poor combustion and flame patterns. Severe cases of tramp air entering the boiler will cause the furnace to go "positive." This creates a hazardous condition known as a blowback, where furnace gases exit the boiler via the service doors rather than the air pollution control device.

The proper delivery of refuse to the furnace grate is an important factor in furnace control. Poor control of refuse feed into the combustion chamber will cause RDF to hit the rear wall of the furnace, resulting in flame and impact erosion at the rear of the boiler, as well as poor combustion patterns on the grates. Also, should there be large concentrations of ground glass in the processed refuse, slagging of the furnace rear wall, side walls, and superheater is likely to occur. This is due to the fact that glass does not burn but melts. The melting glass adheres to the furnace walls and is cooled by the furnace draft. This was a serious problem with the "first generation" processing/burn facilities that were not equipped with glass removal devices. Many operating facilities have been retrofitted with trommels and or disc screening systems for the removal of glass. This has led to a marked reduction in clinkering of the furnace.

RDF contains halogenated hydrocarbons that, when burned, produce corrosive compounds such as hydrogen chloride, hydrogen fluoride, and chlorine. As previously stated, the only metal that will effectively deter corrosion from these compounds is titanium, which is expensive. Thus, proper furnace control must be used in order to keep waterwall tube wastage at a minimum. In addition, because of suspension burning in a dedicated unit, the amount of fly ash generated is greater than in a mass burning unit. The fly ash is highly erosive, causing tube wear damage in the superheater and economizing tube banks. Silicon carbide (SiC) in both the waterwall and the superheaters has aided in deterring tube wastage. The Dade County, Florida cogeneration facility initially experienced severe superheater wastage due to the erosive and corrosive effects of the fly ash/hot gas combination. However, the management has since decided to install SiC refractory on its superheater tubes. Results have been reported to be positive by Resources Recovery (Dade County), Inc.

The aforementioned problems and shutdowns of RDF facilities have led to a general uneasiness within industries, utilities, investment banks, insurance firms, and communities to pursue RDF as a viable means for resource recovery. Consequently, these types of facilities may be more difficult to finance and insure than waterwall plants.

6
SMALL-SCALE SYSTEM ECONOMICS

INTRODUCTION

The economics of converting solid waste to a fuel form such as refuse-derived fuel (RDF) or an energy form such as steam includes both direct and indirect expenditures associated with a waste-to-energy project. This chapter highlights the economics of three different waste-to-energy technologies and compares these to a conventional sanitary landfill disposal system. The methodology for determining the economics of small-scale waste disposal systems, including a comparison of sanitary landfill versus waste-to-energy system, is presented in this chapter. This methodology is valid to determine project economics and initial feasibility. However, prior to project financing, a much more detailed investigation will be required to assess the most current cost and revenue components and the sensitivities of costs over the project life.

The capital costs of a waste-to-energy facility are dependent on factors such as the selected technology, air pollution control equipment, site-specific and subsurface conditions, and the requirements or needs for ancillary support buildings for waste processing, administrative activity, and/or vehicle maintenance. Factors such as the method of financing and the front-end planning costs that may be financed also impact the capital costs to be financed. The operations and maintenance costs include labor, utilities, maintenance, insurance, residue haul and disposal, and facility management costs. Facility revenues are derived from the sale of energy (steam and/or electricity) and the sale of recovered material, from tipping fees, and to a lesser extent, from earnings on reserve funds, i.e., debt service reserve.

The major cost and revenue elements of a revenue producing solid waste disposal technique (i.e., waste-to-energy facilities) are depicted in Figure 6-1. The total cost of the disposal facility is a function of the capital cost annualized as the debt service payment (the annual mortgage) plus the annual operation and maintenance (O&M) costs. Once a sanitary landfill or a waste-to-energy system has been built, the total capital costs are set and usually become an annually budgeted line item reflective of the total principal and interest payments. (This is similar to the mortgage on a home or a fixed-term lease on an automobile.) Just as a home may be an average family's largest

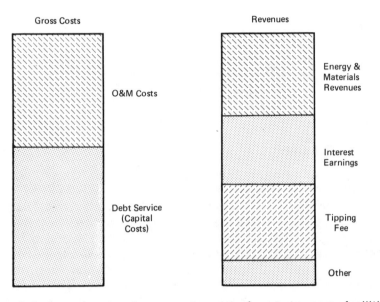

Figure 6–1. Annual cost and revenue elements of waste-to-energy facilities.

investment, most waste disposal systems that are being built today are also the local community's most capital intensive program for long-term waste disposal. In this context, long-term usually varies from 15 to 25 years for solid waste projects and the costs are in the millions.

THE BASE CASE—CONVENTIONAL WASTE MANAGEMENT SYSTEMS

Solid waste management system costs are typically classified as collection, hauling, and final disposal (i.e., landfill) costs. These cost components apply to both municipal and industrial waste management systems. These three components are usually obvious in municipal systems. An industrial manager, however, sometimes overlooks the internalized collection costs of delivering the waste from the work sites where it is generated to a bin receptacle which is then transported to a central loading dock or container.

Regardless of whether the waste is municipal, industrial, or institutional in nature, the breakdown of the three cost components for typical full-service municipal solid waste systems reflects that more than 75 percent of the total disposal system cost is collection related. Of the remaining 25 percent, hauling costs per ton are usually somewhat more than the final disposal costs per ton. In other words, only about 10 to 15 percent of the total waste management costs relate to final disposal.

It is important to note that while much attention has been devoted to the rapid escalation in final disposal (landfill) costs, the actual landfill cost component (i.e., the disposal operation) really represents a relatively minor component of a community's total waste management program costs. An inefficient collection system, which could represent more than 75 percent of the total waste management costs, is usually unaffected by the construction of a new waste-to-energy facility. A new system would only affect the haul costs if route travel distances were altered. Present and future total costs for existing solid waste management programs based on landfilling are expected to increase significantly. These projected increases reflect the effects of compliance with the Resource Conservation and Recovery Act of 1976 (RCRA), the increasing remoteness of approved disposal sites, and the inflation-based increases in labor, trucks, and fuel for hauling.

While one may take issue with the specific projections, it is clear that a landfill-based waste management system will experience steadily rising costs without producing revenue offsets through the sale of recovered products. The capital and operating costs for a series of sanitary landfill sizes have been estimated to provide a data base for comparing the economics of the conventional system with the waste-to-energy options. These detailed capital costs for landfill development are provided in Table 6–1, with the estimated annual O&M costs detailed in Table 6–2.

The capital costs for developing a new sanitary landfill system in a size range of 50 to 500 TPD (tons per day) are expected to range from $1.25 to $5.62 million (1983).

In Tables 6–1 and 6–2, the footnotes provide a rough outline of the key assumptions used in the calculations, with the recognition that these are "generic costs" for these types of facilities. There will obviously be local site-specific situations with topography, soils, equipment availability, land costs, etc., that could significantly alter specific assumptions considered herein.

Based on the cost data presented, a new 50 TPD landfill would be expected to cost about $22.20 per ton in 1983, with the economies of scale reducing the cost of a 500 TPD landfill to $12.60 per ton.

TRANSPORTATION NETWORK

The transportation network established for delivering waste to a landfill or to a waste-to-energy facility is an integral part of the development process and the costs associated with such a project. Public acceptance of the project

Table 6-1. Sanitary Landfill System Capital Costs (1983 Dollars).

	TONS PER DAY			
	50	100	200	500
I. *Capital Cost*				
Site Preparation[a]	250,000	300,000	400,000	800,000
Buildings[b]	80,000	150,000	300,000	500,000
Utilities	50,000	100,000	150,000	150,000
Equipment[c]	300,000	350,000	600,000	1,400,000
Liner[d]	100,000	200,000	250,000	400,000
Leachate Control System	50,000	100,000	170,000	400,000
Groundwater Monitoring (incl. wells)	10,000	25,000	40,000	45,000
Subtotal	840,000	1,225,000	1,910,000	3,695,000
Land[e]	200,000	500,000	750,000	1,000,000
Contingency Fund (10% subtotal)	84,000	122,500	191,000	369,500
Legal, Detailed Site Engineering, and Administration (15% subtotal)	126,000	183,750	286,500	554,250
TOTAL	1,250,000	2,031,250	3,137,500	5,618,750
Annual Cost[f]	170,200	305,800	481,000	909,000

[a] Includes road construction, fencing, clearing and grubbing, screening, excavation, and grading.
[b] Includes personnel and equipment buildings plus scales and appurtenances.
[c] Equipment costs include:

	TONS PER DAY			
	50	100	200	500
Tractor (Crawler) Bulldozers	(1)	(1)	(2)	(3)
Scraper	(1)	(1)	(1)	(2)
Steel Wheel Compactor	(1)	(1)	(1)	(1)
Water Truck	—	—	(1)	(1)
Pickup	—	(1)	(1)	(1)

[d] Five year liner costs were calculated at $0.41/sq ft of working cell bottom surface. Additional liner will be purchased during 5th, 10th, and 15th year of operation.
[e] Land valued at $5,000/acre.
[f] Equipment costs were amortized over 5 years (50 TPD—10 years) at 11%, with development costs amortized at 11% for 20 years.

will be impacted by routing changes and perceived ease of delivery. In addition, commercial haulers will need to be thoroughly briefed on routing changes to minimize nonproductive time during transit.

Vehicle routing for solid waste disposal options is also of concern to residents and businesses which could be impacted by an increase in traffic volume on their streets or variations in vehicle types. A proposed project site located some distance from the population centroid may be unacceptable due to the long haul distance. In the same manner, a landfill or waste-to-

Table 6–2. Sanitary Landfill System O&M Costs (1983 Dollars).

	TONS PER DAY			
	50	100	200	500
II. *Annual O&M Costs*				
Labor[a]	60,000	85,000	130,000	300,000
Equipment O&M[b]	30,000	62,000	75,000	200,000
Utilities	3,000	8,000	10,000	15,000
Cover Material[c]	20,000	50,000	70,000	85,000
Groundwater Monitoring	12,000	20,000	25,000	50,000
Administration[d]	6,000	9,500	13,000	30,000
Equipment Replacement Fund[e]	40,000	110,000	175,000	360,000
Cell Development	5,000	10,000	15,000	25,000
Subtotal	176,000	354,500	513,000	1,055,000
O&M Cost ($/ton)	11.28	11.36	8.22	6.76
III. *Total Annual Cost*[f]	346,000	660,300	994,000	1,964,000
Unit Cost ($/ton)[g]	22.20	21.20	15.90	12.60

[a] Labor includes:

	TONS PER DAY			
	50	100	200	500
Senior Engineer	—	—	—	(1)
Foreman/Supervisor	(1)	(1)	(1)	(1)
Mechanics	—	(1)	(1)	(1)
Equipment Operator	(1)	(2)	(2)	(3)
Laborers	(1)	(1)	(1)	(2)
Scale Operator/Secretary	—	—	(1)	(2)
Total	3	5	6	10

With 40% overhead and benefits.
[b] Includes fuel, oil, lubrication, tires, and repairs.
[c] Assumes 50/50 blending with on-site excavated material.
[d] 10% of labor costs.
[e] One-fifth (one-tenth for 50 TPD) of present value of equipment and liner. Equipment assumed to be replaced every 5 years. Interest earned on replacement fund will offset escalation of equipment prices.
[f] Includes annual cost for capital from Table 6–1.
[g] Unit costs based on 6 days/week operation (312 days/year).

energy site close to the population center may be equally undesirable due to potential disruption of a residential neighborhood or existing traffic patterns.

Because of the significance of a project's transportation network, an analysis of the transportation impacts in terms of ton-miles, total miles traveled, and annual costs is usually undertaken for new landfills and waste-to-energy scenarios. A baseline case representing the existing waste disposal practices is generally used for comparison with the waste-to-energy scenarios.

Existing Refuse Collection System

The existing municipal or private refuse collection system would be assessed to determine the number of commercial and residential routes. Vehicles collecting waste along these routes would be identified.

If a multijurisdictional system is to be implemented, outside collection districts and all private haulers licensed to collect refuse would be identified. However, since collection may not be mandatory, vehicle routing could be highly dependent upon the demand for contract refuse disposal. Licensed haulers would be interviewed for this analysis and can provide useful input on hauling routes and landfill destinations.

For small businesses and private citizens disposing of their own waste, routing is a matter of driver preference and, typically, proximity to the existing landfills open for disposal, regardless of origin.

Analysis Parameters

The analysis should evaluate the haul distances and impacts in terms of ton-miles, total traveled miles, and annual costs for the existing waste disposal transportation system and the waste-to-energy scenarios.

For comparison and evaluation purposes, basic parameters should be established to standardize the scenarios. Typical parameters and the method for calculating ton-miles, total traveled miles, and annual costs are presented here.

To provide a basis for distributing the waste stream, the study area would be divided into waste sheds. This is done by correlating the information provided by public or privately licensed haulers and 1980 U.S. Bureau of Census census tracts. Where possible the hauler-identified routes and district boundaries are followed. However, in cases of conflicts, the census tract boundaries may be used.

A waste shed centroid is defined as that point within the waste shed which represents an equal distribution of the population concentration and transportation links that would commonly be traveled by loaded waste vehicles. These centroids are used to calculate haul distances from the waste sheds to the disposal points.

Haul routes are planned over major arterials and connector streets to avoid disruption of residential neighborhoods and major changes in vehicle patterns. If the major local transportation corridor ran east and west, north-south routing could require traveling along secondary commercial and residential streets with limited opportunities to avoid known congested areas. This would result in increased haul times which need to be reflected in a ton-mile calculation. Other routing considerations include avoiding commercial strips that tend to be congested due to numerous stoplights and intersections. When possible, routing should be on state highways. Primary access routes should be identified for each scenario analyzed.

Calculation of Haul Distances

Once the waste stream generation and general routing assumptions have been established, haul distances can be calculated for transporting waste to a disposal point. Solid waste in a study area may be hauled directly from the generation point to one or several disposal sites. The transportation of waste from the generation source to a disposal site is defined as the direct haul segment of the transportation network.

The locations of the existing operating landfills may or may not be convenient for most residents. As local landfills are closed, waste disposal may be consolidated at a future landfill or waste-to-energy facility. Therefore, parts of the region could have greater "direct haul" distances. The result of a substantial increase in the "direct haul" is a loss of the sense of convenient disposal for residents and a quantifiable increase in haul costs.

In certain rural areas, large percentages of the waste stream may be hauled directly by private citizens, so providing convenient disposal in the populated areas becomes integral to securing the waste stream. The inconvenience of increased direct haul distances must be determined by analyzing the transportation grid. Based on experience, the cutoff point for economic direct haul is around 15 miles one way. As the distance from the waste generation areas to the disposal site increases, it becomes economical to consolidate the waste by building a transfer station as an intermediate collection point. Here, wastes from many smaller collection vehicles are transferred to fewer, more efficient, large-capacity trailers. This significantly reduces that portion of the waste disposal system's costs associated with transporting waste. However, this reduction in haul costs must be considered in light of the increased cost of building and operating the transfer stations.

For evaluation purposes, haul distances and the corresponding annual costs can be broken down into two components:

1. direct haul from collection route to intermediate disposal (transfer station) or final disposal site, and
2. transfer haul from transfer station to final disposal site.

Direct haul distances and, where appropriate, transfer haul distances need to be developed. The cost of direct haul would be borne by the private resident or commercial haulers (assumed to be a pass-through to the client) contracting with the waste generators and not calculated as part of the overall system economics. Depending upon the configuration of the overall transportation network, direct haul could be either to a transfer station or to a point of final disposal such as a landfill or energy recovery facility. Transfer haul refers to the transporting of waste collected at the transfer station to the landfill or waste-to-energy facility. Transfer haul is typically accomplished by tractor-trailer rigs hauling 65 cubic yard trailers. This cost is borne by the transfer station owner/operator and is reflected in the station tipping fees.

Transfer stations may be sized to accommodate the present quantity of waste or may also provide for future increases in waste quantities. If necessary, additional tractors and trailers can be added at any time to existing stations to increase capacity. Such an addition would also increase the number of hauls from the station.

Residue Disposal

A landfill is an integral part of any waste-to-energy facility and provides a disposal site for unprocessible materials and process residue. The transport of these materials from the facility is an important component of the transportation network, and the associated costs must be included in the analysis. The residue haul distances from each of the proposed scenarios must be developed.

Additional Considerations

Certain components of the transportation network are affected by unique characteristics of the transportation grid. Bridge restrictions need to be identified, and routing to and from the proposed project sites and waste sheds will be impacted in terms of distances and costs by routing restrictions of this nature. In general, the scenarios under consideration must allow some flexibility in vehicle routing which minimizes the need to upgrade existing streets or to construct new access routes.

Transportation Cost Analysis

A transfer station will not lower the door-to-door collection cost or the disposal cost. Savings are realized only by reducing the haul distance from the collection zone to the unloading area. Because collection trucks travel only short distances to unload at a transfer station, they can be back on their routes while a transfer vehicle containing several collection truckloads is traveling to a distant disposal site.

Although a transfer station operation offers potential savings, it requires an extra materials handling step and the construction of a transfer facility. The associated costs must be recovered (i.e., refuse collection/disposal fees increased) or money will be lost in the transfer operation. The costs that are incurred are

- Capital expenditures for land, structures, and equipment
- Labor, utilities, maintenance, operating, and overhead costs at the transfer plant
- Labor, operating, maintenance, and overhead costs incurred in the bulk hauling operation

Costs are saved with the utilization of a transfer operation because

- The nonproductive labor time is cut since collectors no longer ride to and from the disposal site. Therefore, the larger the collection crew, the greater is the savings.
- Any reduction in mileage traveled by the collection trucks results in a savings in operating costs, and in addition, it may be possible to reduce the number of collection crews needed because of increasing the productive collection time.

The final step in the evaluation of the transportation network is to develop costs for each haul component.

Transportation System Cost Estimation Procedure

The methodology used in estimating direct hauling costs can also be applied to the transfer system. The method involves estimating the total "time-based" costs and "mileage-based" costs of the transfer haul and converting these figures to a total "cost per minute." To this transfer haul cost must be added the nontransport costs of owning and operating the transfer station and the cost of maneuvering and unloading the transfer vehicle. These nontransport costs will be converted to a "cost per ton" basis for comparative purposes.

A hypothetical example is presented in order to demonstrate the procedure for estimating the costs of a transfer system. A medium-capacity, 11 cubic yard stationary hydraulic compactor transfer station with a rated capacity of 300 TPD is used for illustrative purposes. Although the design capacity is 300 TPD, this example assumes the actual loading to be 200 tons of solid waste per day. The difference is accounted for by expected future growth. The plant operates on a 5 day per week, 8 hour workday schedule, which means there are 260 operating days per year. Personnel requirements include

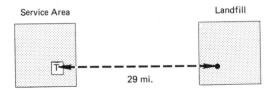

$\boxed{T}$= Location of transfer station

Figure 6-2. Discharge site (landfill or waste-to-energy plant).

three employees (an operations supervisor, a compactor operator, and a weighmaster). A total of seven 65 cubic yard transfer trailers and three diesel tractors is required to transport the "present" volume of refuse to the disposal site located 29 miles from the facility, with two transfer trailers kept as backups. Figure 6-2 shows the location of the transfer facility relative to the discharge site. It has been estimated that a transfer trailer requires a total round trip time of about 110 minutes to travel to and from the discharge site, including the time to unload at the site.

The steps used in the cost estimating procedure are outlined in Table 6-3, with data pertaining to the hypothetical example included to illustrate the

Table 6-3. Cost Estimating Procedure.

STEP 1. ESTIMATION OF TOTAL TIME-BASED TRANSPORTATION
COST PER MINUTE

1. Calculate annual amortization cost of tractor trailer:

Tractor cost	$55,000
Trailer cost	35,000
Vehicle cost	$90,000

Assumed 5 year life and 12% annual interest rate:

$$\frac{\text{Cost (\$)}}{90,000} \times .2774 \text{ (CR factor)}^{b} \qquad = \qquad \frac{\text{ANNUAL COST (\$)}}{25,000}$$

2. Equipment redundancy and/or spares @ 25%:   6,250

3. Determine driver's salary and fringe benefits (consult municipal records):

$$\frac{\text{Salary (\$)}}{24,000} + \frac{\text{Fringe Benefits (\$)}}{6,000 \text{ (est. 25\%)}} \qquad = \qquad 30,000$$

4. Estimate vehicle insurance, licenses, taxes (consult municipal records or local insurance agencies):   3,000

Table 6–3. (continued)

5. Calculate total time-based transportation cost per year: $64,250

6. Calculate total time-based transportation cost per minute (Assumed 5 day work week, 8 hour workday):

$64,250/yr ÷ 124,800 min/yr   = $0.515/min

STEP II. ESTIMATION OF TOTAL MILEAGE-BASED TRANSPORTATION COST PER MINUTE

1. Calculate mileage-based transportation cost per mile for fuel:

$1.30/gal ÷ 4 mi/gal   = $0.325/mi

2. Calculate mileage-based transportation costs of $2,000/year for tires and oil:

$2,000/vehicle/yr ÷ 17,400 vehicle-mi/yr[a]   = $0.115/mi

3. Calculate mileage-based transportation costs per mile for maintenance and repair:

$3,500/vehicle/yr ÷ 17,400 vehicle-mi/yr   = $0.201/mi

4. Calculate total mileage-based transportation costs: $0.641/mi

5. Calculate total mileage-based transportation costs per minute:

$0.64 (total cost/mi) × 58 (round trip mi) ÷ 110 (min. per round trip) = $0.338 (cost/min)

---

[a] Mileage calculated from transfer station to disposal site to transfer station (15,000 mi/yr) and includes 15% additional mileage for off-route travel.

STEP III. ESTIMATION OF TOTAL TRANSPORTATION COST PER TON PER MINUTE

1. Calculate the total transportation cost per minute:

$0.515/min + $0.338/min   = $0.85/min

2. Calculate the total transportation cost per ton per minute:

$0.85/min ÷ 15 tons/vehicle[a]   = $0.06

---

[a] Diesel tractor trailer assumed to have 15 ton payload capacity in order not to exceed highway weight limits. It should be noted that this may result in less than full loads for the 65 cu yd transfer trailers.

[b] Capital recovery factor.

STEP IV. ESTIMATION OF TOTAL COST OF OWNING, OPERATING, AND MAINTAINING THE TRANSFER FACILITY

1. Estimate the total annual cost of owning, operating, and maintaining the transfer facility (excluding the cost of transfer trailers and their operation):

(cont.)

Table 6–3. (continued)

STEP IV. (cont.)	
Station and equipment cost	$ 900,000
Engineering and construction management costs	150,000
Hydraulic compactor cost	150,000
Total Capital Cost	$1,200,000
Labor cost (supervisor, compactor operator, weighmaster)	$ 80,000
Compactor maintenance	10,000
Power	5,000
Utilities	500
Office supplies	1,000
Scale maintenance	2,000
Insurance	1,500
Property maintenance	1,500
Total Annual Operating Cost	$ 101,500

Final capital cost to be amortized over 20 years at 12%:

$1,200,000 (capital cost) × .1339 (UCR factor) = $160,600 (annual amortization)

2. Add annual capital cost to annual operating and maintenance cost to determine total annual cost of transfer station: $160,600 + 101,500 = $262,100.

3. Calculate annual cost per ton of owning and operating transfer station:
$262,100 (annual cost) ÷ 200 tons/day ÷ 260 operating days/yr = $5.04/ton

4. Calculate maneuvering and unloading cost per ton at landfill site:
10 min to unload × $0.515/min ÷ 15 tons/vehicle = $0.34 (cost/ton)

---

costing methodology. It should be noted that even though these costs are unique to this particular example, the methodology can be used to estimate the costs of owning and operating a transfer station for other situations.

To summarize, the calculated estimated transportation total cost from Table 6–3 for the transfer system was $0.06 per ton per minute. The total fixed cost for owning and operating the transfer facility was $5.04 per ton. The fixed cost of unloading and maneuvering the transfer vehicles was $0.34 per ton. Thus, the total fixed cost of the transfer operation is $5.38 per ton ($5.04 + $0.34 = $5.38). A graph for the transfer haul cost can be plotted, and the graph for this example transfer station is shown in Figure 6–3.

WASTE-TO-ENERGY PROJECT COSTS

Today, many communities are faced with solid waste management problems and, in particular, the difficult decision of implementing a waste-to-energy system versus continuing their traditional disposal of solid waste through landfilling. The choice usually becomes one of economics, but this is not always the case. For many municipalities, and this number is increasing, there is no landfill alternative or perhaps no reasonable landfill option. The

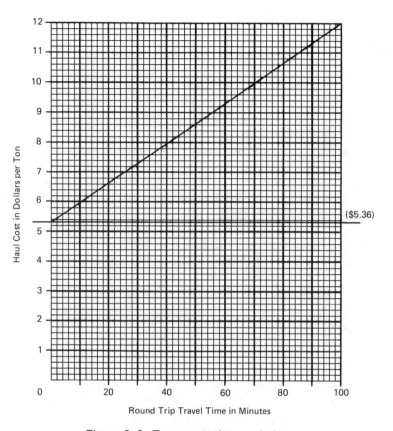

Figure 6-3. Transportation worksheet.

availability of suitable land may be limited or there may be a host of competing land uses. More likely, the ability to site a new sanitary landfill in proximity to the area of waste generation, or even remote from that area, may be obviated by overwhelming public opposition due to concern over the perceived environmental effects.

The "not in my back yard" cry has become very common to solid waste disposal planning and development, and is not restricted to the siting of landfills. Developers of waste-to-energy projects are confronting obstacles in the form of organized public protest just as formidable, and this has had, and is still having, a negative impact on project implementation in several locales.

While the environmental implications of new solid waste facilities have become a primary consideration in the decision-making process, the first question still voiced by most public officials is, "How much is it going to cost?"

The focus of this chapter is on that question and the factors that affect the answer. It seems like a simple question, but it is really one that is rather complex and leads to other issues that need to be understood and analyzed. For example, the capital cost of one alternative versus another may be significant; however, the life cycle cost may be quite comparable or even significantly lower for the alternative that presents the higher "first" costs.

Costs should also be considered on a time-value basis or in the context of present worth. The projected cost of solid waste disposal in the year 2000 may be chilling when the effects of inflation are considered, but viewing the costs based on their present value in current dollars may offer a different perspective. Municipal officials and project planners may overlook the present value cost implications of a project. This orientation could lead to a decision for an alternative that, over its life cycle, will actually be more costly to the community in terms of the time value of expenditures made or savings realized.

There are many factors that affect the cost of waste-to-energy versus landfill and the time frame when the costs of each become equivalent. Perhaps the most significant factor is the market for energy to be produced by a waste-to-energy project. A community may have a strong desire to implement a waste-to-energy project and may possess the other essential building blocks (i.e., site, front-end resources, sufficient waste supply, strong leadership, etc.) needed to develop a project, but it may lack a market for the energy or, more likely, a market whose fossil fuel costs, energy demand, location, future plans, and/or stability are compatible with the development of a successful project. For example, a market in a stable industry using No. 2 oil as a fuel source and having a reasonably consistent steam demand compatible with the potential energy that could be generated from the local waste supply could be expected to offer the opportunity for better project life cycle economics than a market with seasonal energy demand fluctuations that uses coal or a combination of coal and natural gas as its boiler fuel. Too, a utility that is experiencing load growth and having a significant portion of its base load generating capacity provided by oil-fired units is likely to offer a better opportunity for higher electricity prices to a project than one that is base loaded by coal, nuclear, and/or hydro capacity; has significant excess capacity; and is forecasting limited need for new capacity over the life cycle of the waste-to-energy project.

Tables 6–4 and 6–5 show the value of municipal solid waste when converted to steam and displacing different fossil fuels. From the tables, it can be seen that the type of fossil fuel displaced has a substantial impact on the steam value and would significantly affect waste-to-energy project economics.

Table 6–4. Relative Costs of Fuels.

FUEL COSTS (APRIL 1984 $)	FUEL ($/10⁶ BTU)	FUEL VALUE[a] ($/1,000 LB OF SATURATED STEAM)
Fuel Oil		
No. 2 ($0.98/gal)	7.00	8.75
No. 6 ($0.69/gal)	4.55	5.69
Natural Gas	4.06	5.07
($4.26/1,000 cu ft)		
Coal ($40/ton)	1.75	2.19

[a]Based on 80% boiler efficiency of existing facilities.

While there are several other factors that affect the value of steam, the tables are presented to show the relative impact of the fossil fuel value to project revenues and, hence, project economics.

It should be noted that several waste-to-energy projects built in the late 1970s and supplying steam to industrial markets that rely on No. 2 oil to raise steam have not realized the energy sales revenues of preconstruction feasibility projections due to the softening of fossil fuel prices and, particularly, oil prices. Concurrently, with the cooling of inflation, facility operating costs may also have lowered, but the spread between the percentage of annual operating cost increase and energy price increase is now very narrow at best. Sensitivity analyses may reveal the magnitude of impact on project costs if the annual inflation rate of oil prices outstrips the rate of general inflation in the economy and continues to over the life cycle of the waste-to-energy facilities presented.

There are also special locational or environmental factors which may singly, or in combination, create economic impacts that tend to favor one alternative over another. A long steam line needed to serve a market may add

Table 6–5. Fuel and Product Values by Fuels Displaced.

FUEL DISPLACED	FUEL COSTS	($/10⁶ BTU)	FUEL[a] VALUE ($/1,000 LB STEAM)	AVERAGE PRODUCT[b] FUEL VALUE ($/TON MSW)
Fuel Oil No. 2	$0.98/gal	7.00	8.75	33.47
Fuel Oil No. 6	$0.69/gal	4.55	5.69	21.76
Natural Gas	$4.26/1,000 cu ft	4.06	5.07	19.39
Coal	$40/ton	1.75	2.19	8.38

[a]Based on 80% boiler efficiency of existing facilities.
[b]Based on 4,500 lb of steam per ton MSW and a 15% market discount.

several million dollars to a project. For a small-scale facility, this steam line could increase the project cost 25 percent or more. Additional air pollution control equipment, if required to control acid gases, could raise the facility capital cost substantially and increase the annual owning and operating cost about $5 to $10 per input ton.

On the other hand, the costs for construction and operation of a new state-of-the-art sanitary landfill which includes a special liner (even a double liner as required now in some states), a leachate collection and treatment system, sufficient rolling stock to operate it properly, and reserves for closure and long-term maintenance following closure are also likely to be substantial. If a community must develop or use a landfill at a location remote from its waste generation centroid, the incremental costs to haul or perhaps to develop a transfer station and transfer haul solid waste (as noted earlier in this chapter) could also be considerable.

The manner in which a project is financed has an important bearing on its initial costs as well as life cycle costs. For a capital intensive waste-to-energy project, it is usually necessary to issue bonds to finance all or the major portion of project costs. Currently, there is a trend in financing such projects with tax-exempt revenue bonds (see Chapter 7). This often includes a private equity contribution as well and project implementation on a "full-service" basis. As part of the financing, there may be special reserves, capitalized interest, insurance, and other costs that add markedly to the cost of the facility.

A landfill, by comparison, is usually less capital intensive than a waste-to-energy plant (for the same refuse tonnage handled) and is frequently financed through general obligation debt if publicly owned, by a loan, or even out of current government revenues. In many cases, the land may already be owned by the municipality so the cost is strictly for development and equipment. Often, it is a matter of expanding an existing landfill, either horizontally or vertically or both. The comparison is not necessarily one of a new landfill versus a waste-to-energy facility but rather one of an expanded, albeit upgraded, landfill for all community waste versus a waste-to-energy facility and a modified landfill for residue, bypass waste, and non-processibles. The waste-to-energy plant may be the most capital intensive public works project ever developed in some communities and, as such, may engender concern before community officials fully consider the life cycle economic impact.

CAPITAL COSTS

Capital costs include all costs associated with design, construction, and start-up of a solid waste management facility. Capital cost elements include site development, buildings, equipment, utility, construction management, start-up, acceptance testing, and administration costs. Land costs are included if land is purchased for the project. (If land is leased, it would be included with the annual operating costs.) The capital costs also include all costs for steam piping to the steam customer or electricity tie-in equipment to the utility grid. Additional capital costs are also incurred to obtain the necessary project permits and to pay design, legal, and other consultant fees. A listing and brief description of the major capital cost elements are as follows:

- Design
- Land acquisition and site development
- Buildings
- Waste handling/loading equipment
- Combustion/steam generation equipment
- Power generation equipment (if an electricity generating project)
- Steam/condensate transmission line(s)
- Electrical switchgear and transmission line (if appropriate)
- Air pollution control equipment
- Stack(s) (including foundation and erection)
- Spare parts
- Bid performance and payment bonds
- Insurance
- Start-up and initial operations tests
- Acceptance tests

Site Development.   Site development costs include costs for clearing and grubbing, excavation and grading, tie-in of utilities, fencing, lighting, and landscaping. Site development costs vary and are contingent on subsurface and topographical conditions of the site.

Buildings.   Buildings are required for the scale-house, tipping floor, boiler enclosures, vehicle maintenance areas, waste processing areas, control rooms, and administrative functions. Building costs are site specific and are also influenced by the communities or towns in which the waste-to-energy facility is located. Building costs are relatively lower in the Sunbelt states as the boilers and turbines do not require enclosures. Some communities may insist on additional expenditures to acquire an aesthetically pleasing building for their neighborhood.

Equipment.   Equipment costs are usually reported as installed costs. However, to estimate costs, it is important to note if the costs are FOB manufacturing plant, delivered to site, or installed. Plant mechanical and electrical costs may be shown as two individual cost elements. Rolling stock and vehicles are usually included as a separate line item. Rolling stock includes front-end loaders, pickup trucks, maintenance vehicles, and sweepers.

Energy Market Tie-in. All steam pipeline costs, condensate return line costs, and electrical switchgear costs should be included in the capital costs. Underground or overhead mode of transmission of the steam lines or electrical feed lines will impact the overall cost.

Acceptance Testing. On completion of facility construction, the system vendor must verify that his equipment meets all negotiated performance guarantees. In most cases, this also includes testing for air emissions to determine if the combustion process meets all state permit air emission requirements. Start-up costs and a facility testing reserve fund should be capitalized to provide operating funds during the start-up period before full revenue payments begin.

Legal and Administrative Costs. Costs for legal and administrative services prior to, and through, acceptance testing should be included in the capital costs.

Engineering and Contingency Costs. Costs for design engineering and construction management are usually estimated at 10 percent of the total installed costs. Contingency fees of 10 to 20 percent of the total installed cost are included in the overall capital cost estimate. The contingency cost estimate is a function of the level of detail that went into the estimate and supporting cost engineering review.

Possible additions to the above-mentioned capital costs could include:

- Community development expenses
- Special insurance
    Efficacy insurance
    Bond insurance
    Title insurance
    Other
- Sales tax
- Ancillary facilities
- Trustee expenses

OPERATIONS AND MAINTENANCE COSTS

The annual O&M costs include such items as labor, utilities, repair and replacement of parts, supplies, insurance, residue disposal, and bypassed waste disposal costs. The annual budget of existing small-scale direct combustion systems employing modular incinerator technology typically varies from 30 to 40 percent spent on debt service payments to about 60 to 70 percent spent on O&M costs. The O&M costs gradually rise, based on the inflation rate, to become an even larger portion of the annual costs in later years of the project.

The O&M costs are itemized as follows:

- Labor (base wages, fringe benefits, and overtime allowance)
- Utilities (waste, sewage disposal, electricity, gas, fuel oil)
- Energy plant maintenance and supplies
- Waste receiving/handling equipment maintenance and supplies
- Building maintenance and supplies
- Raw materials
- Contract services
- Equipment rental
- Site lease (if any)
- Equipment replacement/maintenance fund
- Insurance
- Residue haul and disposal
- Fuel for maintenance vehicles

In addition to these, there may be project-specific O&M costs:

- Special fees and/or taxes
    Host community fee(s)
    Payment(s) in lieu of taxes
    Management/operating fees
    Letter of credit fees
    Inspection fees
    Fees for special bonding/insurance to satisfy energy customer or municipality
    Trustee fees
- Residue processing
- Transport of recovered materials
- Administrative (plant specific as well as municipality specific)
- Legal/engineering/other due to change in environmental compliance requirements
- Disposal of bypassed wastes

All of these itemized O&M costs will vary depending on the technology selected and the size of the facility. The labor requirements depend on the number of employees per shift and the number of shifts operated per day. The incinerator ash or other residue generation rates will depend on the type of waste processing or combustion facility. Most waste combustion systems are equipped with a wet ash system; therefore, the wet weight (not "dry residue" weight) is used for estimating hauling and disposal costs.

Most small waste-to-energy systems consume power on the order of 50 kilowatt-hours per ton of incoming waste and, if a waste-to-electricity facility, sell the net power to an electricity market. Therefore, for determination of revenues, the "net" steam (pounds per hour) and "new electricity" (kilowatt-hours) to be sold must be considered (unless it is planned to operate in a sell-all, buy-all electricity generation system where the costs versus revenues will balance each other out in the final cost analysis).

Other incidental or unexpected O&M costs could result from faulty facility equipment, labor strikes, explosion damage to buildings or equipment, or some unforeseen mechanical or electrical breakdown.

Based on the design capacity of the facility, its on-line availability, scheduled maintenance, and variations in waste delivery rates, it is possible that at certain times of the year the facility (due to limited refuse storage capacity) may have to direct some solid waste to the landfill. The O&M costs should account for the hauling and transportation costs of this bypassed waste component.

In order to offer perspective on the relative life cycle cost impact of developing a waste-to-energy facility as compared to a new state-of-the art landfill of approximately the same capacity, cost projections were prepared for several sizes of facilities as well as for different waste-to-energy technologies.

Tables 6–6 through 6–17 contain examples of a 50 TPD_7[1] modular incinerator, a 100 TPD_7 modular incinerator, a 200 TPD_7 modular incinerator, a 250 TPD_7 waterwall incinerator, a 700 TPD_5 refuse-derived fuel processing plant, and a 380 TPD_7 RDF dedicated boiler system. *All costs are generic and are intended only to show the relative economics on a comparative basis given certain assumptions. We have not attempted to show the cost effects of differing levels of risk allocation, varying levels of performance guarantees or security enhancements, and alternative ownership arrangements.* Obviously, as discussed earlier, there is a plethora of variables and conditions specific to a given project that will impact the final costs. Moreover, the

---

1. TPD_7 refers to the daily waste processing capacity on a seven day basis expressed in tons per day. TPD_5 refers to the waste processing capacity on a five day basis expressed in tons per day.

Table 6-6. 50 TPD$_7$ Modular Incineration: Installed Capital Costs (1985 Dollars).

COST COMPONENTS	CAPITAL COSTS ($1,000) STEAM ONLY	
I. General Construction		
A. Land (2 acres)[a]	30	
B. Site Preparation/Building	710	
C. Mechanical/Electrical Installation	330	
Subtotal, Item I		1,070
II. Equipment		1,730
III. Contractor's Overhead and Profit[b]		220
IV. Contingencies and Omissions[c]		450
V. Fees and Construction Costs		
A. Design and Construction Management[d]	170	
B. Legal and Project Management[e]	70	
C. Management/Financial Consultants[f]	50	
D. Start-up Costs[g]	120	
E. Contingencies on Fees and Construction Costs[h]	20	
Subtotal, Item V		430
Total		3,900
($/ton of capacity)	78,000	

[a]Land value: $15,000 per acre.
[b]Based on 8% of IB, IC, and II costs.
[c]Includes insurance and loading costs during construction and construction administration. Cost based on 15% of Items, I, II, and III.
[d]Based on 5% of Items I, II, III, and IV.
[e]Based on 2% of Items I, II, III, and IV.
[f]Based on 1.5% of Items I, II, III, and IV.
[g]Based on 3 months of annual O&M costs.
[h]Based on 5% of Items VA through VD

Table 6-7. 50 TPD$_7$ Modular Incineration: Operating and Maintenance Costs (1985 Dollars)

COST COMPONENTS	O&M COSTS ($1,000) STEAM
Labor[a]	225
Utilities[b]	70
Maintenance, Parts, and Supplies[c]	45
Services[d]	10
Equipment Replacement[e]	20

(cont.)

Table 6-7. (continued)

COST COMPONENTS	O&M COSTS ($1,000) STEAM
Insurance[f]	10
Residue Disposal[g]	80
Total	460
Cost per Ton ($)	31.55

[a]Total number of employees for the steam plant would be 10. Labor rates would range fron $15,000 to $30,000 per year, which includes 25% for fringe benefits. Average salary assumed to be $22,500 including fringe benefits.
[b]Utilities include electricity, water/sewer, and fuel for the on-site rolling stock.
[c]The cost per ton of solid waste incinerated under the steam option for this line item would be $3.00 per ton based on the annual throughput of the incineration system at 14,600 tons.
[d]Services include accounting, legal, and miscellaneous (e.g., laundry, telephone).
[e]Based on 1% of the capital cost for equipment.
[f]Based on 0.5% of the capital cost for general construction and equipment.
[g]Includes charges for hauling and disposal. Hauling costs were based on $0.50 per ton-mile. Annual residue would be 4,380 tons, or 30% of the processed waste. The round trip haul distance was set at 15 miles. The disposal charge at the landfill is estimated at $10.00 per ton.

Table 6-8. 100 TPD, Modular Incineration: Installed Capital Costs (1985 Dollars).

COST COMPONENTS	CAPITAL COSTS ($1,000)		
	STEAM	COGENERATION	ELECTRICITY
I. General Construction			
A. Land (3 acres)[a]	50	50	50
B. Site Preparation/Building	1,200	1,300	1,300
C. Mechanical/Electrical Installation	655	764	764
Subtotal, Item I	1,905	2,114	2,114
II. Equipment	2,828	3,724	3,910
III. Contractor's Overhead and Profit[b]	378	467	482
IV. Contingencies and Omissions[c]	767	946	976
V. Fees and Construction Costs			
A. Design and Construction Management[d]	294	362	374
B. Legal and Project Management[e]	117	145	150
C. Management/Financial Consultants[f]	88	109	112
D. Start-up Costs[g]	190	220	240

Table 6–8. (continued)

COST COMPONENTS	CAPITAL COSTS ($1,000)		
	STEAM	COGENERATION	ELECTRICITY
E. Contingencies on Fees and Construction Costs[h]	35	42	44
Subtotal, Item VI	724	878	920
Total	6,602	8,129	8,402
($/ton of capacity)	66,020	81,290	84,020

[a] Land value: $15,000 per acre.
[b] Based on 8% of IB, IC, AND II costs.
[c] Includes insurance and loading costs during construction and construction administration. Cost based on 15% of Items I, II, and III.
[d] Based on 5% of Items I, II, III, and IV.
[e] Based on 2% of Items I, II, III, and IV.
[f] Based on 1.5% of Items I, II, III, and IV.
[g] Based on 3 months of annual O&M costs.
[h] Based on 5% of Items VA and VD.

Table 6–9. 100 TPD$_7$ Modular Incineration: Operating and Maintenance Costs (1985 Dollars).

COST COMPONENTS	O&M COSTS ($1,000)		
	STEAM	COGENERATION	ELECTRICITY
Labor[a]	325	425	475
Utilities[b]	100	110	110
Maintenance, Parts, and Supplies[c]	90	110	115
Services[d]	20	20	20
Equipment Replacement[e]	30	40	40
Insurance[f]	20	30	30
Residue Disposal[g]	155	155	155
Total	740	890	945
Cost per Ton ($)	25.35	30.50	32.35

[a] Total number of employees for the steam, cogeneration, and electricity options would be 13, 17, and 19, respectively. Labor rates would range from $15,000 to $35,000 per year, which includes 25% for fringe benefits. Average salary assumed to be $25,000 including fringe benefits.
[b] Utilities include electricity, water/sewer, and fuel for the on-site rolling stock.
[c] The cost per ton of solid waste incinerated under the steam, cogeneration, and electricity options for this line item would be $3.00, $3.75, and $4.00, respectively, based on the annual throughput of the incineration system at 29,200 tons.
[d] Services include accounting, legal, and miscellaneous (e.g., laundry, telephone).
[e] Based on 1% of the capital cost for equipment.
[f] Based on 0.5% of the capital cost for general construction and equipment.
[g] Includes charges for hauling and disposal. Hauling costs were based on $0.50 per ton-mile. Annual residue would be 8,760 tons, or 30% of the processed waste. The round trip haul distance was set at 15 miles. The disposal charge at the landfill is estimated at $10.00 per ton.

Table 6–10. 200 TPD₇ Modular Incineration:
Installed Capital Costs (1985 Dollars).

COST COMPONENTS	CAPITAL COSTS ($1,000)		
	STEAM	COGENERATION	ELECTRICITY
I. General Construction			
A. Land (4 acres)[a]	60	60	60
B. Site Preparation/Building	1970	2080	2080
C. Mechanical/Electrical Installation	870	1092	1090
Subtotal, Item I	2,900	3,230	3,230
II. Equipment	5,630	7,230	7,560
III. Contractor's Overhead and Profit[b]	680	840	860
IV. Contingencies and Omissions[c]	1,380	1,690	1,750
V. Fees and Construction Costs			
A. Design and Construction Management[d]	530	650	670
B. Legal and Project Management[e]	210	260	270
C. Management/Financial Consultants[f]	160	190	200
D. Start-up Costs[g]	300	340	350
E. Contingencies on Fees and Construction Costs[h]	60	70	70
Subtotal, Item V	1,260	1,510	1,560
Total	11,850	14,500	14,960
($/ton of capacity)	59,250	72,500	74,800

[a]Land value: $15,000 per acre.
[b]Based on 8% of IB, IC, AND II costs.
[c]Includes insurance and loading costs during construction and construction administration. Cost based on 15% of Items I, II, and III.
[d]Based on 5% of Items I, II, III, and IV.
[e]Based on 2% of Items I, II, III, and IV.
[f]Based on 1.5% of Items I, II, III, and IV.
[g]Based on 3 months of annual O&M costs.
[h]Based on 5% of Items VA and VD.

Table 6–11. 200 TPD₇ Modular Incineration:
Operating and Maintenance Costs (1985 Dollars).

COST COMPONENTS	O&M COSTS ($1,000)		
	STEAM	COGENERATION	ELECTRICITY
Labor[a]	425	450	500
Utilities[b]	180	200	200
Maintenance, Parts, and Supplies[c]	180	220	230
Services[d]	20	20	20
Equipment Replacement[e]	60	80	80

Table 6–11. (continued)

COST COMPONENTS	O&M COSTS ($1,000)		
	STEAM	COGENERATION	ELECTRICITY
Insurance[f]	40	60	60
Residue Disposal[g]	310	310	310
Total	1,215	1,340	1,400
Cost per Ton ($)	20.80	22.95	24.00

[a] Total number of employees for the steam, cogeneration, and electricity options would be 17, 18, and 20, respectively. Labor rates would range from $15,000 to $35,000 per year, which includes 25% for fringe benefits. Average salary assumed to be $25,000 including fringe benefits.
[b] Utilities include electricity, water/sewer, and fuel for the on-site rolling stock.
[c] The cost per ton of solid waste incinerated under the steam, cogeneration, and electricity options for this line item would be $3.00, $3.75, and $4.00, respectively, based on the annual throughput of the incineration system at 58,400 tons.
[d] Services include accounting, legal, and miscellaneous (e.g., laundry, telephone).
[e] Based on 1% of the capital cost for equipment.
[f] Based on 0.5% of the capital cost for general construction and equipment.
[g] Includes charges for hauling and disposal. Hauling costs were based on $0.50 per ton-mile. Annual residue would be 21,900 tons, or 30% of the processed waste. The round trip haul distance was set at 15 miles. The disposal charge at the landfill is estimated at $10.00 per ton.

Table 6–12. 250 TPD, Waterwall Incineration: Installed Capital Costs (1985 Dollars).

COST COMPONENTS	CAPITAL COSTS ($1,000)		
	STEAM	COGENERATION	ELECTRICITY
I. General Construction			
A. Land (5 acres)[a]	80	80	80
B. Site Preparation/Building	2,840	2,950	2,950
C. Mechanical/Electrical Installation	1,530	1,860	1,860
Subtotal, Item I	4,450	4,890	4,890
II. Equipment	7,830	8,940	9,650
III. Contractor's Overhead and Profit[b]	1,840	2,070	2,180
IV. Contingencies and Omissions[c]	2,120	2,390	2,510
V. Fees and Construction Costs			
A. Design and Construction Management[d]	810	910	960
B. Legal and Project Management[e]	330	370	380
C. Management/Financial Consultants[f]	240	270	290
D. Start-up Costs[g]	390	440	450

(cont.)

Table 6–12. (continued)

	CAPITAL COSTS ($1,000)		
COST COMPONENTS	STEAM	COGENERATION	ELECTRICITY
E. Contingencies on Fees and Construction Costs[h]	90	100	100
Subtotal, Item V	1,860	2,090	2,180
Total	18,100	20,380	21,410
($/ton of capacity)	72,400	81,520	85,640

[a] Land value: $15,000 per acre.
[b] Based on 8% of IB, IC, AND II costs.
[c] Includes insurance and loading costs during construction and construction administration. Cost based on 15% of Items I, II, and III.
[d] Based on 5% of Items I, II, III, and IV.
[e] Based on 2% of Items I, II, III, and IV.
[f] Based on 1.5% of Items I, II, III, and IV.
[g] Based on 3 months of annual O&M costs.
[h] Based on 5% of Items VA and VD.

Table 6–13. 250 TPD₇ Waterfall Incineration: Operating and Maintenance Costs (1985 Dollars).

	O&M COSTS ($1,000)		
COST COMPONENTS	STEAM	COGENERATION	ELECTRICITY
Labor[a]	575	675	700
Utilities[b]	200	220	220
Maintenance, Parts, and Supplies[c]	240	290	310
Services[d]	30	30	30
Equipment Replacement[e]	80	90	100
Insurance[f]	60	70	70
Residue Disposal[g]	380	380	380
Total	1,565	1,755	1,810
Cost per Ton ($)	21.45	24.05	24.80

[a] Total number of employees for the steam, cogeneration, and electricity options would be 23, 27, and 28, respectively. Labor rates would range from $15,000 to $40,000 per year, which includes 25% for fringe benefits. Average salary assumed to be $25,000 including fringe benefits.
[b] Utilities include electricity, water/sewer, and fuel for the on-site rolling stock.
[c] The cost per ton of solid waste incinerated under the steam, cogeneration, and electricity options for this line item would be $3.25, $4.00, and $4.25, respectively. Annual cost is based on the annual throughput of the incineration system at 73,000 tons.
[d] Services include accounting, legal, and miscellaneous (e.g., laundry, telephone).
[e] Based on 1% of the capital cost for equipment.
[f] Based on 0.5% of the capital cost for general construction and equipment.
[g] Includes charges for hauling and disposal. Hauling costs were based on $0.50 per ton-mile. Annual residue would be 21,900 tons, or 30% of the processed waste. The round trip haul distance was set at 15 miles. The disposal charge at the landfill would be $10.00 per ton.

Table 6–14. 700 TPD$_5$ Refuse-Derived Fuel Processing Plant: Installed Capital Costs (1985 Dollars).

COST COMPONENTS	CAPITAL COSTS[a] ($1,000)
I. General Construction	
A. Land (10–15 acres)[b]	150
B. Site Preparation/Building	3,820
C. Mechanical/Electrical Installation	1,310
Subtotal, Item I	5,280
II. Equipment[cde]	3,820
III. Contractor's Overhead and Profit	1,370
IV. Contingencies and Omissions	1,570
V. Fees and Construction Costs	
A. Design and Construction Management	840
B. Legal and Project Management[f]	240
C. Management/Financial Consultants[g]	180
D. Start-up Costs[h]	390
E. Contingencies on Fees and Construction Costs[i]	80
Subtotal, Item V	1,730
Total	13,770
($/ton of capacity)	19,670

[a] Dollar values in the table are rounded.
[b] Land value: $15,000 per acre.
[c] Processing equipment includes: trommel screens, shredders, air classifiers, and magnetic separators.
[d] Auxiliary equipment includes: RDF storage, dust control, loadout equipment, electrical controls, and instrumentation.
[e] Rolling stock includes: front-end loaders and tractor/trailers for residue hauling.
[f] Based on 15% of Items IB, IC, components of II.
[g] Based on 15% of Items I, II, and III.
[h] Based on 8% for one year of Items I, II, III, and IV.
[i] Based on 8% of Items I, II, III, and IV.

Table 6–15. 700 TPD$_5$ Refuse-Derived Fuel Processing Plant Operating and Maintenance Costs (1985 Dollars)

COST COMPONENTS	O&M COSTS ($1,000)
Labor[a]	450
Utilities and Fuel[b]	200
Maintenance, Parts, and Supplies[c]	290
Services[d]	30
Equipment Replacement[e]	40
Insurance[f]	40

(*cont.*)

Table 6–15. (continued)

COST COMPONENTS	O&M COSTS ($1,000)
Residue Disposal[g]	500
Total	1,550
Cost per Ton ($)	10.80

[a]Total number of employees would be 18. Labor rates would range fron $15,000 to $40,000 per year, which includes 25% for fringe benefits. Average salary assumed to be $25,000 including fringe benefits.
[b]Utilities include electricity, water/sewer, and fuel for the rolling stock and plant heat.
[c]Based on $2.00 per ton of solid waste processed by the processing plant (144,000 tons).
[d]Services include accounting, legal, and miscellaneous (e.g., laundry, telephone).
[e]Based on 1% of the capital cost for equipment.
[f]Based on 0.5% of the capital cost for general construction and equipment.
[g]Includes charges for hauling and disposal of plant residue estimated at 20% of processed stream (exclusive of ferrous recovery product). Hauling costs were based on $0.50 per ton-mile. The annual residue, which includes the noncombustible/nonferrous components of the processed waste would be 28,800 tons. The round trip haul distance was set at 15 miles. The disposal charge at the landfill would be $10.00 per ton.

Table 6–16. 380 TPD₋RDF Dedicated Bioler: Installed Capital Costs (1985 Dollars).

COST COMPONENTS	CAPITAL COSTS[a] ($1,000)		
	STEAM	COGENERATION	ELECTRICITY
I. General Construction			
A. Land (3 acres)[b]	—	—	—
B. Site Preparation/Building	870	980	980
C. Mechanical/Electrical Installation	2,400	2,730	2,730
Subtotal, Item I	3,270	3,710	3,710
II. Equipment[cde]	9,860	11,610	12,480
III. Contractor's Overhead and Profit	1,970	2,300	2,430
IV. Contingencies and Omissions	2,270	2,640	2,790
V. Fees and Construction Costs			
A. Design and Construction Management	870	1,010	1,070
B. Legal and Project Management	350	410	430
C. Management/Financial Consultants[f]	260	300	320
D. Start-up Costs[g]	380	440	480

Table 6–16. (continued)

COST COMPONENTS	CAPITAL COSTS[a] ($1,000)		
	STEAM	COGENERATION	ELECTRICITY
E. Contingencies on Fees and Construction Costs[h]	90	110	120
Subtotal, Item V	1,950	2,270	2,420
Total	19,320	22,530	23,830
($/ton of capacity)	50,840	59,290	62,710

[a] Dollar values in the table are rounded.
[b] Land owned by project developers.
[c] Processing equipment includes: trommel screens, shredders, air classifiers, and magnetic separators.
[d] Auxiliary equipment includes: RDF storage, dust control, loadout equipment, electrical controls, and instrumentation.
[e] Rolling stock includes: front-end loaders and tractor/trailers for residue hauling.
[f] Based on 15% of Items IB, IC, components of II.
[g] Based on 15% of Items I, II, and III.
[h] Based on 8% for one year of Items I, II, III, and IV.

Table 6–17. 380 TPD, RDF Dedicated Boiler System: Operating and Maintenance Costs (1985 Dollars).[a]

COST COMPONENTS	O&M COSTS ($1,000)		
	STEAM	COGENERATION	ELECTRICITY
Labor[b]	390	520	630
Utilities[c]	550	600	600
Maintenance, Parts, and Supplies[d]	280	360	390
Services[e]	40	40	40
Equipment Replacement[f]	90	110	110
Insurance[g]	60	70	80
Residue Disposal[h]	230	230	230
Total	1,640	1,930	2,080
Cost per Ton MSW ($)[i]	11.38	13.40	14.45

[a] Source: Gershman, Brickner ;a& Bratton, Inc.
[b] Total number of employees for the steam, cogeneration, and electricity options would be 14, 19, and 23, respectively. Labor rates would range from $15,000 to $40,000 per year, which includes 25% for fringe benefits. Average salary assumed to be $27,500 including fringe benefits.
[c] Utilities include water/sewer, electricity, and fuel for the on-site rolling stock.
[d] The cost per ton of RDF processed under the steam, cogeneration, and electricity options would be $2.50, $3.25, and $3.50, respectively. The annual throughput of the boiler is 111,000 tons of RDF.
[e] Services include accounting, legal, and miscellaneous (e.g., laundry, telephone).
[f] Based on 1% of the capital cost for equipment.
[g] Based on 0.5% of the capital cost for general construction and equipment.
[h] Includes charges for hauling and disposal of 12% of RDF as residue. Hauling costs were based on $0.50 per ton-mile. Annual residue would be 13,300 tons, which includes ash and RDF bypassed to landfill when boiler system is down. The round trip haul distance was set at 15 miles. The disposal charge at the landfill would be $10.00 per ton.
[i] Total annual tonnage (as MSW) would be 144,000 tons before RDF production.

ultimate decision as to which alternative to implement may not be driven by costs, and with the waste-to-energy alternative, an acceptable landfill will still be necessary as part of the overall project.

WASTE-TO-ENERGY PROJECT ECONOMICS

In assessing the potential costs of a waste-to-energy system, the key thing that must always be remembered is that no two projects are the same. Even if the technical parameters of two plants are similar, the site topography, climate, soil conditions, local construction costs, state of the general economy, or a change in energy sales conditions and other factors could affect final project costs. Two plants could be identical, but financing them 12 months apart could lead to significant interest rate changes that also would affect the comparable economics of the projects.

The possible disparity in project economics from facility to facility begins with the decisions that are made in the very early planning stages of the project. When decisions are made that move the project risks and responsibilities from the public sector to the private sector, cost increases will undoubtedly occur. The level of cost increases must be commensurate with the actual risks versus those perceived risks that the private sector has been asked to assume. The following pages present an overview of capital and operating costs associated with different small-scale waste-to-energy projects that have been constructed. Depending upon the type of procurement followed, the risks and roles of the public versus private sector groups will also affect costs and the project schedule.

It is possible to have private contractors operating or owning and operating the facilities. If the private contractor only operates the facility, it would typically charge a management fee on the order of 10 percent of O&M costs and perhaps require a share of product sales revenues as a performance incentive. If the contractor contributes equity and assumes ownership and operating responsibilities along with their attendant risks, it would likely require a greater share of product sales revenues or cash flow from the project to provide a sufficient return on equity beyond the tax benefits of ownership that would be available.

In reviewing potential capital and O&M cost data, the following tables provide a review of the cost of plants that have been in operation for a few years as well as new facilities being developed at the time of this writing:

Table 6–18: Capital Costs—Modular Incineration Plants
Table 6–19: Capital Costs—Waterwall Incineration Plants
Table 6–20: Capital Costs—Refractory Incineration Plants
Table 6–21: Capital Costs—RDF Dedicated Boiler Plants

Table 6-18. Capital Cost per Installed Ton[a]— Modular Incineration Plants.

FACILITY LOCATION	START-UP	DESIGN CAPACITY (TPD)[b]	CAPITAL COST ($ MILLION)	INSTALLED COST PER TON ($)
Older Plants				
Pittsfield, MA	1981	240 (s)	10.85	45,200
Auburn, ME	1981	200 (s)	4.40	22,000
Portsmouth, NH	1982	200 (s)	6.20	31,000
Cattaraugus Co., NY	1983	112.5 (s)	5.40	48,000
Tuscaloosa, AL	1984	300 (s)	8.30	27,500
Plants under Construction or Recently Financed				
Pascagoula, MS	1984	150 (s)	5.90	39,330
Oswego Co., NY	1985	200 (c-g)	16.00	80,000
Cleburne, TX	1985	100 (c-g)	8.50	85,000

[a]Does not include financing costs.
[b]s = steam; c-g = cogeneration.

As can be seen from these tables, the technology of the plant, as well as the plant size and year of start-up, all affect the plant's capital costs. However, the impact of the project's capital costs must be reduced to a more meaningful and comparable number such as the capital cost on a dollar per ton

Table 6-19. Capital Cost per Installed Ton[a] Waterwall Incineration Plants.

FACILITY LOCATION	START-UP	DESIGN CAPACITY (TPD)	CAPITAL COST ($ MILLION)	INSTALLED COST PER TON ($)
Pinellas Co., FL	1983	2,000	103.0	51,500
Westchester Co., NY	1984	2,250	178.9	79,500
Baltimore, MD	1984	2,250	185.0	82,200
North Andover, MA	1985	1,500	123.0	82,000
Older Plants				
Hampton, VA	1980	200	10.4	52,000
Gallatin, TN	1982	200	12.4	62,000
New Hanover Co., NC	1984	200	13.8	69,000
Plants under Construction or Recently Financed				
Claremont, NH	1985	200	17.7	88,500
Marion Co., OR[b]	1987	550	47.5	86,360
Dutchess Co., NY	1987	400	30.5	76,250

[a]Does not include financing costs.
[b]Includes acid gas control equipment.

**Table 6-20. Capital Cost per Installed Ton[a]—
Refractory Incineration Plants.**

FACILITY LOCATION	START-UP	DESIGN CAPACITY (TPD)	CAPITAL COST ($ MILLION)	INSTALLED COST PER TON ($)
Harrisonburg, VA	1982	100	9.2	92,000
Glen Cove, NY[b]	1983	250	24.0	96,000

[a]Does not include financing costs.
[b]Codisposal of sewage sludge with MSW.

processed basis. Table 6-22 presents all four technologies previously discussed and adds the financing costs as well as certain other project-specific costs that were included in each project's actual financing plan. Therefore, the start-up dates, type of technology, and plant configuration can be reviewed for the capital cost or debt service impact on the project.

In addition to the capital costs of the plant, the actual "bottom line" cost to the users must incorporate the plant's operating and maintenance costs (O&M) as well as the actual or projected revenues from the energy or materials recovery steps. Once again, the historical data base is one customary reference that can be tapped to elicit a general range of values. Table 6-23 presents the O&M costs and revenue data from several plants around the country. It can be noted that significant and widely varied O&M as well as revenue values are reflected in the table. Major entries that affect the O&M variance include labor costs (union versus nonunion shops), local tax payments or payments-in-lieu of taxes (PILOTs), reserve and contingency

**Table 6-21. Capital Cost per Installed Ton—
RDF Dedicated Boiler Plants.[a]**

FACILITY LOCATION	START-UP	DESIGN CAPACITY (TPD)	CAPITAL COST ($ MILLION)	INSTALLED COST PER TON ($)
Niagara Falls, NY	1981	2,250 MSW/ 1,900 RDF	165	69,300
Albany, NY	1981	800 MSW/750 RDF	155	37,500
Dade County, FL	1982	3,000 MSW/ 2,250 RDF	30	55,000
Haverhill, MA	1984	1,300 MSW/ 850 RDF	58	44,800

[a]At the time this book went to press, no new small-scale RDF dedicated boiler plants had been financed or constructed.

Table 6–22. Summary of Project Costs.

FACILITY LOCATION	START-UP	CAPACITY (TPD)	ENERGY FORM	CAPITAL COST ($ MILLION)	TOTAL COST ($ MILLION)	DEBT SERVICE ($/TON)
Mass Burning—Modular Incinerators						
Auburn, ME	1981	200	Steam	4.40	4.90	11.12
Pittsfield, MA	1981	240	Steam	10.85	12.35	12.10
Portsmouth, NH	1982	200	Steam	6.20	6.36	18.40
Mass Burning—Waterwall Incinerators						
Hampton, VA	1980	200	Steam	10.40	10.4	9.10
Gallatin, TN	1982	200	Cogeneration	12.44	15.9	35.77
New Hanover Co., NC	1984	200	Cogeneration	13.84	14.0	39.38
Mass Burning—Refractory Furnaces						
Harrisonburg, VA	1982	100	Steam	9.20	10.45	50.10
RDF Processing for Co-firing						
Ames, IA	1975	200	RDF	6.82	6.82	12.50
Madison, WI	1979	400	RDF	4.00	4.40	9.11

accounts, and residue hauling/disposal charges. The revenues are most often lowest when coal is the displaced fuel and highest when oil is displaced. The level of the sales discount, or incentive, to the energy purchaser is also important to the final price.

Table 6–23. Incremental Cost and Revenue Review.

FACILITY LOCATION	YEAR	ANNUAL TONNAGE	O&M ($/TON)	REVENUES ($/TON)[a]	TIPPING FEE ($/TON)[b]
Mass Burning—Modular Incinerators					
Auburn, ME	8/82–8/83	48,755	30.16[c]	11.40	29.88
Pittsfield, MA	1983	63,135[d]	19.00 (est.)	20.49[e]	13.25
Portsmouth, NH	7/82–6/83	68,674	20.60	20.75	18.20
Mass Burning—Waterwall Incinerators					
Hampton, VA	1983	76,785	17.66	6.00[f]	7.00
Gallatin, TN	1984	39,000	23.53	18.60	15.10
New Hanover Co., NC	1984	58,400	23.97	20.55	22.70[g]

(cont.)

Table 6–23. (continued)

FACILITY LOCATION	YEAR	ANNUAL TONNAGE	O&M ($/TON)	REVENUES ($/TON)[a]	TIPPING FEE ($/TON)[b]
Mass Burning—Refractory Furnaces					
Harrisonburg, VA	1984	21,900	21.20	27.17	42.33
RDF Processing for Co-firing					
Ames, IA	1982	39,457	22.28	10.88[h]	23.90
Madison, WI	1983	53,274[i]	31.16	6.10	33.19

[a]Includes energy sales only.
[b]Normalized overall tonnage to be received in the designated year. In actuality, there may be different service fees for different users depending on their contract status and the specific "deal" embodied in the project.
[c]Includes $12.96/ton MSW for residue disposal.
[d]Does not include 5,230 tons of paper sludge processed.
[e]Includes energy sales, materials sales, and interest income.
[f]Steam revenues have exceeded debt service and operating costs; therefore, the City contracted with NASA, its steam customer, to pay the costs of owning and operating the plant each month in return for steam at no cost.
[g]Project operation to be subsidized with approximately $1 million in County funds as necessary to reduce tipping fee.
[h]RDF and metal sales only.
[i]Includes 25,336 tons bypassed to landfill. The 27,938 tons processed produced 14,686 tons of RDF.

REVENUES

Relief from the constantly escalating costs for conventional landfills may be achieved through a waste-to-energy system that generates revenues by producing a fuel form or an energy form whose value will increase over time. Capital investments in waste-to-energy facilities must be evaluated not only in terms of the savings associated with replacing a sanitary landfill but also in terms of the projected value of the energy produced from the facility over its designated lifetime.

The steam or electricity sales price is negotiated with the energy customer and is typically based on the relative costs of displaced fuels. Table 6–4, presented earlier, illustrates the product values for displaced fuels. Table 6–5 shows that the value of municipal solid waste (MSW) fuel is directly related to the cost of conventional fuels. Therefore, economics for replacing coal may be marginal, but replacing petroleum-derived fuel with solid waste may be a superior economic alternative. Based on a similar analysis, if electricity if displaced at $0.05 per kilowatt-hour, the value of MSW is estimated at $25.00 per ton.

Revenues are obtained from sale of electricity and steam in cogeneration systems. Depending on the steam market, turbine extracted or exhausted steam may be utilized for heating or process applications. The energy market will determine the size and type of turbine required for a specific project.

Other potential sources of revenues from solid waste processing facilities are the sale of refuse-derived fuel and recovered material products. The refuse-derived fuel can be co-fired with coal or wood, or fired as the sole fuel in dedicated boilers. Based on the location of the boilers and the waste processing facilities, it may be necessary to transport the RDF to the boiler site. Transportation costs of the RDF must be included in the overall economic analysis. The revenue stream of many different projects now in operation was presented in Table 6–23.

LIFE CYCLE COSTS

Life cycle costing is a method for comparing alternatives on a long-term basis. The complexity and longer-term life of waste-to-energy facilities demand this type of an analysis. Rather than comparing only initial capital costs among options, life cycle costing compares differences in total economic impact over the life of the project.

Life cycle costing is especially useful since the costs and revenue streams differ from year to year. The escalation of individual cost and revenue elements may not be at equal rates. Therefore, an economic analysis based on the first year of operation or on the first several years of operation could be misleading. A life cycle analysis incorporates these changing costs and indicates how the cash flows interact to alter annual total costs.

Life cycle costs of various alternatives can be illustrated graphically. Costs can be shown either as remaining relatively constant over time, or as increasing or decreasing in particular periods during the project life. In compiling an analysis of this nature, it is necessary to develop a detailed economic and financing model which sets forth all of the key assumptions in the analysis. An example of the parameters that go into a representative model is presented in Table 6–24. It is important to note that the financing parameters must include the project's construction time frame as well as operations lifetime. The project's technical performance parameters will be used to determine waste inputs and product outputs, and to derive facility revenue stream calculations.

Based upon the evaluation of numerous proposers and bids, as well the performance of countless feasibility studies, the authors have developed a generic curve that provides some rule-of-thumb cost components for quick cost estimating and preliminary analyses. Figure 6–4 presents the capital costs (1984 dollars) for direct combustion systems of the modular incinerator and waterwall incinerator configuration. Two different series of curves reference two vertical cost axes and provide the capital costs in changing from low-pressure steam systems to higher-pressure "all-electric" projects. Figure 6–5 presents the incremental O&M costs for the operation of the

Table 6-24. Typical Economic Analysis Assumptions.

Technology Option to Be Considered:
- Prepared fuel (RDF) dedicated boilers/cogeneration of steam and electricity

Financing Parameters:

• Date of financing (all scenarios)	7/1/85
• Capital cost estimates	7/1/83 dollars
• Escalation period to date of financing	2 years
• Capital cost escalation rate	6%/year
• Construction period (including start-up/acceptance tests)	36 months
• Plant operating period (from 7/1/88)	20 years
• Interest rate on debt (tax-exempt revenue bonds)	10%/year
• Term of capitalized interest	36 months
• Debt service reserve fund	1 year's principal/ interest payment
• Contingency reserve fund	3 months' O&M cost
• Rate of interest earnings on funds during construction period	10%/year
• Private equity contribution of total direct construction costs (includes escalation)	25%
• Bond underwriting fees, legal expenses, and other bond issuance costs (% of total bond issue size)	4%
• Final bond sizing includes escalation during construction	6%/year on balance outstanding

System Parameters:

• Project technology	RDF processing plant/dedicated boiler
• RDF processing plant throughput— MSW (tons/yr)	321,000
• RDF dedicated boiler plant throughput (tons/yr)	281,700
• Design size (tons/day)	Processing $1,400_{5\frac{1}{2}}$[a] Boiler 960_7[b]
• Waste delivery arrangement RDF transfer hauling	10 miles one way
• Energy forms sold	
Steam	X
Electricity	X
• Boiler plant assumptions	
Number of combustion lines	2
Assumed boiler efficiency (%)	72
Internal steam usage (% of generation)	15
Steam pressure/temperature	635 psig/750°F
Total residue to landfill (% MSW)	37.5
RDF combusted (as % of MSW processed)	87.7

Operating/Maintenance Cost Related:
- Facility O&M costs escalate at 6%/year

Table 6-24. (continued)

- Residue haul costs escalate at 6%/year
- Residue/bypass waste disposal costs:
    Base cost—$13.75/ton (7/1/84)
    Escalation @ 6%/year on operations component

Project Revenues (during operations period):
- Escalation rates:
    Natural gas @ 8%/year from 1987 through 2007
    Oil @ 8%/year from 1987 through 2007
    Electricity @ 6%/year
- Earnings on project reserve funds—10%/year

Other Assumptions:
- Front-end project development costs (through construction/start-up/acceptance): included in bond issue, estimated at 1% of the installed capital costs.
- Administrative costs (during 20 year plant operations period): one-half person year with fringes ($18,000).
- Payment in lieu of taxes (PILOT): $1/ton of waste.
- Heating value of MSW is 4,500 Btu/lb; heating value of RDF is 5,000 Btu/lb.

---

[a]Operating week = 5½ days.
[b]Operating week = 7 days.

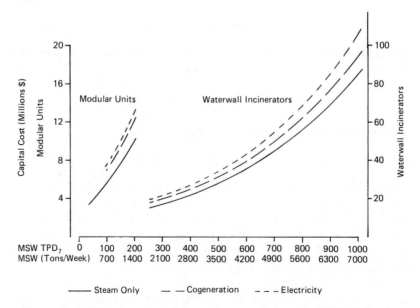

Figure 6-4. Direct combustion system capital costs (1984 dollars).

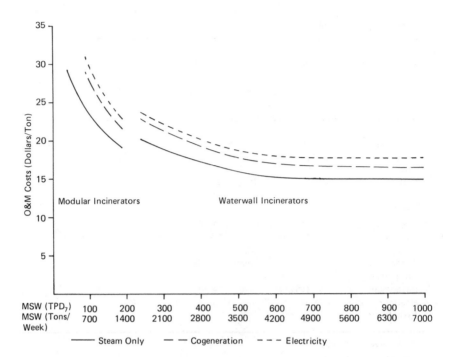

Figure 6-5. O&M costs direct combustion systems (1984 dollars).

plants that were shown in Figure 6-4. In a similar fashion, the presentation of the prepared fuel process technology capital costs in Figure 6-6 and O&M costs in Figure 6-7 affords a cost overview of these technology options.

ECONOMICS OF COGENERATION SYSTEMS

Cogeneration economics are looking much more favorable in many locations throughout the United States where waste-to-energy projects are being pursued. In some locations, electric prices have become so attractive that some projects without good long-term steam markets are considering solely electricity production and not bothering to enter into the more complicated arrangements with industry for cogeneration. More often, steam sales revenues are higher (per Btu generated), but electricity production becomes a desirable adjunct to the base load energy market.

The costs of cogeneration equipment added to waste-to-energy facilities vary widely, depending on the size, configuration, manufacturer, and type

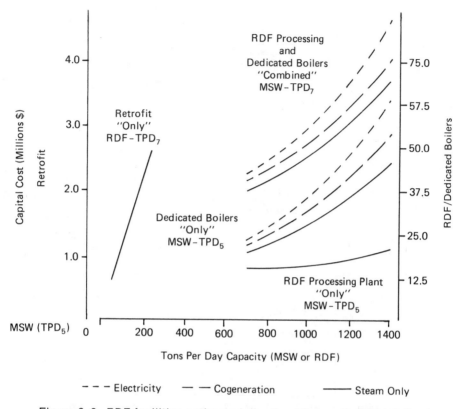

Figure 6-6. RDF facilities estimated direct capital costs (1984 dollars).

of application. However, the incremental costs shown in Tables 6–25 and 6–26 can be used as rough guidelines for the additional capital and O&M costs of adding turbine/generator (T-G) sets to produce electricity. Table 6-25 shows costs of turbine/generator sets for various system back pressure requirements, ranging from condensing applications (at 3 inches of mercury) to 200 psig. As can be seen, the amount of electricity generated per pound of steam input decreases as the back pressure requirement increases (if a constant turbine inlet pressure is assumed). Typically, the capital cost of T-G sets increase as the back pressure decreases (i.e., the lower the exhaust pressure, the higher the number of turbine stages required, which increases costs). Turbine/generators increase in cost with increasing amount of heat recovered from incoming waste. In addition, it can be seen that as the size of the turbine/generator sets increases, with resulting increase in electrical

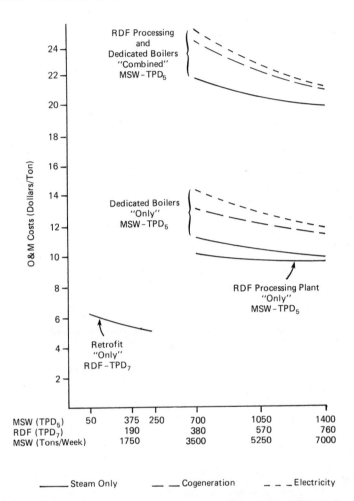

Figure 6–7. O&M costs RDF dedicated boiler systems (1984 dollars).

generation efficiency, the cost impact per ton of solid waste processed decreases substantially. This would be the case for various back pressure configurations. Probably of greater significance is the impact that these turbine/generators have on operating and maintenance costs of the system. These range between $2 and $3 per ton processed and are not as sensitive to the parameters affecting capital cost (see Table 6–26).

Prices that will be offered by utilities for the purchase of electricity depend upon the nature of their current operations. Under the Public Utilities

Table 6–25. Incremental Cogeneration Capital Costs (Turbines, etc.) (1985 dollars).

SYSTEM[a]	CAPITAL COST	INSTALLATION COST	TOTAL COST	ANNUAL COST[b]	IMPACT ON TIPPING FEE ($/TON)
			Modular (100 TPD)		
A	340,000	260,000	600,000	74,000	2.25
B	85,000	64,000	149,000	18,000	0.46
			RDF (600 TPD)		
A	1,500,000	1,000,000	2,500,000	270,000	1.61
B	400,000	270,000	670,000	73,000	0.44
			Mass Burn (500 TPD)		
A	1,200,000	820,000	2,020,000	220,000	1.29
B	400,000	270,000	670,000	73,000	0.43

[a]A = full condensing turbine from 600 psig/750°F to 3 inches Hg.
  B = back pressure turbine from 600 psig/750°F to 150 psig.
[b]Modular systems amortized over 15 years at 9%. All other systems amortized over 20 years at 9%.

Regulatory Policies Act of 1978 (PURPA) "avoided cost" pricing, utilities must offer a price for the electricity which is comparable to the full cost of generating it themselves or purchasing it from another supplier.

Prices of fossil fuels vary widely from state to state. Recently, oil prices ranged from $20.05 per barrel in Arkansas to $41.58 per barrel in Illinois. Gas prices ranged from $1.46 per million cubic feet (mcf) in Florida to $5.10 per million cubic feet in Delaware. Coal prices ranged from $8.46 per ton in North Dakota to $56.02 per ton in Mississippi. Projections of future costs also vary widely according to different assumptions and will result in major

Table 6–26. Incremental Cogeneration Operating Costs (1985 Dollars)

SYSTEM	ANNUAL COST	UNIT COST PER TON
Modular (100 TPD)	99,000	3.00
RDF (600 TPD)	340,000	2.00
Mass Burn (500 TPD)	340,000	2.00

variations in steam and electricity prices negotiated as part of energy market contracts.

As these factors indicate, the revenues for cogeneration projects are highly site specific. In some situations, the waste-to-energy facility would be designed to generate steam at the highest possible pressure and temperature since the total amount of electricity that could be produced is a direct function of the allowable pressure drop through the turbine/generator set. In other situations, steam may be generated at the waste-to-energy technology's maximum pressure and temperature, with the total electricity produced depending on the allowable drop from the turbine inlet steam conditions to the steam conditions (temperature and pressure) required by the energy market.

In many cases, steam will be more attractive to sell than electricity by itself, although the combination of selling steam and electricity in a

Table 6-27. Power Generation Capabilities for Various Mass Burning Process Sizes.

SYSTEM	WASTE RECEIVED WEEKLY (TONS)	MSW[a] BURNED DAILY (TONS)	STEAM FLOW[b] (LBS/HR)	GROSS ELECTRICAL OUTPUT	
				POWER[c] (KW)	ANNUAL[d] (10^6 KWH)
Modular	700	100	17,760	1,300	9.1
	1,400	200	35,800	2,620	18.4
Waterwall	1,400	200	41,050	3,720	26.1
	3,500	500	102,600	10,260	71.9
Refractory	1,400	200	36,220	3,280	23.0
	3,500	500	90,540	9,050	63.4

[a] Waste would be received on a 5.5 day per week basis, but the combustion process would be a 7 day per week operation.
[b] Steam flow based on 4,500 Btu per pound of MSW. Boiler efficiency: modular, 55%; waterwall, 68%; refractory, 60%. Steam conditions: modular, 600 psig, 600°F; waterwall and refractory, 600 psig, 750°F. Steam enthalpy: modular, 1,289.0 Btu per pound of steam; waterwall and refractory, 1,379.5 Btu per pound of steam. Feedwater: 170°F (enthalpy: 137 Btu per pound).
[c] In all cases, outlet pressure: 3 inches Hg via condensing turbine. Other conditions:

	THEORETICAL STEAMING RATE (LBS/KWH)	T-G EFFICIENCY (%)
Modular	8.2	60
Waterwall/Refractory		
200 TPD	7.4	67
500 TPD	7.4	74

[d] Based on an 80% incinerator availability, 365 days per year, 24 hours per day (7,008 hours per year).

cogeneration application most likely will be better for project economics than selling either alone. It should be noted that the feasibility and worthiness of cogeneration are affected by both economic and technical factors. These include

- The relative value of each energy form (steam and electricity) at the specific location under consideration
- The required steam conditions of the project's steam market
- The maximum steam conditions (temperature and pressure) which can be achieved by available waste-to-energy technologies

Although the costs and revenues for cogeneration systems vary widely, typically cogeneration and electricity-only project configurations are more

Table 6–28. Power Generation for Various Mass Burning Process Sizes, Cogeneration.

SYSTEM	WASTE RECEIVED WEEKLY (TONS)	MSW[a] BURNED DAILY (TONS)	STEAM FLOW[b] (LBS/HR)	GROSS ELECTRICAL OUTPUT	
				POWER[c] (KW)	ANNUAL[d] (10^6 KWH)
Modular	700	100	17,900	300	2.1
	1,750	200	35,800	590	4.1
Waterwall	1,750	200	41,050	900	6.3
	3,500	500	102,600	2,240	15.7
Refractory	1,750	200	36,220	790	5.5
	3,500	500	90,540	2,240	15.7

[a] Waste would be received on a 5.5 day per week basis, but the combustion process would be a 7 day per week operation.

[b] Steam flow based on 4,500 Btu per pound of MSW. Boiler efficiency: modular, 55%; waterwall 68%; refractory, 60%. Steam conditions: modular, 600 psig, 600°F; waterwall and refractory, 600 psig, 750°F. Steam enthalpy: modular, 1,289.0 Btu per pound of steam; waterwall and refractory, 1,379.5 Btu per pound of steam. Feedwater: 170°F (enthalpy: 137 Btu per pound).

[c] In all cases, outlet pressure: 150 psig, saturated via back pressure turbine. Other conditions:

	THEORETICAL STEAMING RATE (LBS/KWH)	T-G EFFICIENCY (%)
Modular	28.4	47
Waterwall/Refractory		
200 TPD	23.8	52
500 TPD	23.8	59

[d] Based on an 80% incinerator availability, 365 days per year, 24 hours per day (7,008 hours per year).

expensive than good low-pressure, steam-only configurations. However, there will be situations in which the relative impact of price shifts in the different energy recovery scenarios is important. With only relatively small changes in price conditions, the tipping fees for cogeneration will be affected dramatically. This reinforces the fact that negotiating a strong energy market agreement or combination of agreements is essential to achieving favorable tipping fees. With the increasing prices for electricity as a result of PURPA and other factors, it is important to explore cogeneration opportunities, although they are clearly not going to be advantageous in every situation. Table 6–27 and 6–28 present more specific technical assumptions and design parameters for plant sizing in cogeneration and "electricity-only" modes.

7
FINANCING ALTERNATIVES FOR SMALL-SCALE SOLID WASTE-TO-ENERGY PROJECTS

INTRODUCTION

The financing of a waste-to-energy facility typically entails a long and difficult planning process. The actual financing, i.e., the sale of bonds, is done over a short period; however, the activities and planning leading up to this step take many months and many turns. Structuring financing which is acceptable to all parties-at-interest in the project and, most important, will be attractive to the sources of project capital, typically the bondholders, requires careful planning and substantial input from experienced financial advisors such as investment bankers who have arranged successful waste-to-energy plant financings. Such financial experts should be involved early in the project to examine alternative financial structures and develop a sound financing plan.

Waste-to-energy projects generally have not been received favorably by the investment community. They are perceived as capital intensive, new, and risky, even though certain waste-to-energy technologies have been widely applied for years in Europe, Japan, and elsewhere. Waste-to-energy projects financed with revenue bonds have had limited acceptance in the financial market, and the issues have been characterized by high yields and high debt service coverage requirements. Investors are generally not risk takers. They desire security and liquidity in their investments. In view of the perceived risks, many investors see better investment opportunities in other sectors of the tax-exempt bond market.

Most small-scale waste-to-energy projects (i.e., those in the 100–500 TPD range) operating today were financed with general obligation bonds at interest rates of 5–$6\frac{1}{2}$ percent. Some projects even received government grants and/or loans which further enhanced the financing and kept debt service costs down. Several medium-scale and large-scale projects financed between 1972 and 1980 included project financing with the use of revenue bonds. Interest rates ranged from 6 to $9\frac{1}{2}$ percent and averaged approximately 100 basis points (one percentage point) higher than the Bond Buyers Index (BBI) for other tax-exempt bonds at the time of issue. The most recent of these projects also included additional security such as the taxing power of a local or state government, municipal bond insurance, and/or increased reserve requirements to enhance their financeability.

There have been few small-scale projects developed on a "project financing" basis as private ownership deals. Until recently, most of the vendors in the small-scale marketplace have had limited financial resources to apply as vendor equity or to pledge for financial guarantees necessary to secure, in part, a project financing that incorporates revenue bonds. Projects to be financed in the next several years will face more security requirements than those earlier projects developed largely in the 1970s. During that period, the risks in waste-to-energy were still not well understood by project developers and investors, and there were fewer examples of nonperforming facilities to point to.

Waste-to-energy project financing is now in a critical period. The next generation of projects will place heavy demands on capital resources over the next several years. They will compete with many other types of projects, often considered more secure investments, in the capital markets. Many of these facilities will be the largest single public works project ever undertaken in a number of communities. General obligation bond financing of such projects may no longer be viable in communities with limited remaining debt capacity and several competing demands for capital. This is particularly true in the wake of Proposition 13 in California and Proposition $2\frac{1}{2}$ in Massachusetts where caps on municipal spending have decreased the capability of local governments to put their full faith and credit behind a financing. The availability of government grants and loans, which supported some earlier projects, has decreased substantially in recent years. Thus, there will be fewer opportunities to incorporate such "low-cost" capital in the financing. Further, changing tax laws and the uncertainty of future legislative actions affecting tax benefits could reduce investor interest.

Yet, in the face of the various existing and potential impediments to waste-to-energy project financing, there are some security features, developer incentives, and creative financing mechanisms being brought to bear which may serve to enhance financing in the near-term. The application of more and broader insurance coverage has already been mentioned. The insurance industry is pioneering new forms of coverage with higher limits. Stronger equipment performance guarantees are being made as technology matures and vendor experience broadens. Too, mergers, acquisitions, and joint venture arrangements are emerging in the marketplace which have served to strengthen and expand the limited financing resources and financial guarantees of several small-scale system vendors. Finally, the investment banking industry has sought to enhance project financing, much like the home mortgage banking industry has sought to create new forms of home financing, through innovative financing techniques which make project economics more attractive in the earlier years when revenues are expected to be lower. Financing programs which include adjustable rate bonds, deferred equity, and revenue stabilization are now being integrated with the more conventional project financing structures. In any case, there is still a trend towards project financing which includes some form of tax-exempt revenue bonds. In many projects, this will undoubtedly be the only realistic approach, with greater care given to assuring sufficient cash flows and reserves to cover bonded indebtedness.

The next section examines the different types of financing that are typically considered in a waste-to-energy project and presents the critical elements and characteristics of each type.

TYPES OF TAX-EXEMPT FINANCING AVAILABLE TO WASTE-TO-ENERGY FACILITIES

Introduction

Several alternatives are available for financing waste-to-energy projects. Each project, depending on its location, design, vendor, procurement arrangement, and development agency, presents a different set of possibilities. Further complicating the selection of financing is the complexity of projects. It is often necessary to finance components other than the waste-to-energy facility, e.g., transfer station, residue landfill, ancillary equipment, or modifications at an energy market. These components may be financed by one mechanism and the waste-to-energy plant via another. Combinations of financing arrangements may also be applied in segmented projects whereby a portion of the project is publicly financed while the remainder is privately financed.

One must seek to identify the approach most suitable to the community needs. Each project has its own set of circumstances which is why one cannot recommend a "best" form of financing for every waste-to-energy project. The goal in any financing is to structure the financing to produce a readily marketable security and, at the same time, accomplish the economic needs of all project participants. It is important that all participants have sufficient economic incentives to keep their commitments to the project. Following is a description of the various financing options that local governments and project development agencies may consider.

General Obligation Bond Financing

The simplest financing approach for waste-to-energy project development is the issuance of general obligation (G.O.) bonds. Such bonds are backed by the full faith and credit of the issuing municipality(ies). The municipality(ies) pledges its taxing power without limit as to rate or amount to ensure payment of the debt rather than relying only on project revenues. For a community with a favorable rating, the financial community does not examine G.O. bonds as closely as other financing instruments. Thus, securing financing without certain key project elements in place, e.g., long-term market agreements, is possible. In general obligation bond financing, the municipality is normally required by law to secure voter approval for the bond issue. Thus, the project may come under close public scrutiny if not close examination from the investors.

In cases where a municipality is at or near its legal debt limit or has a poor credit rating, G.O. bonds may not be a viable option for waste-to-energy project financing. Generally, however, G.O. bond financing is the simplest instrument to execute, provides the lowest interest rate and finance charges, and should always be considered as an option. An added advantage is the limited investment assistance necessary in structuring a G.O. bond compared to other options with which a community is usually less familiar.

The credit rating of the municipality determines the salability and price of the bonds. The rating agencies examine such factors as debt limits, local economy and revenue base, and past bond repayment history in assigning a rating. Investors are generally willing to accept a lower return on a higher-rated bond. These bonds are typically intermediate- to long-term securities maturing in 10 to 20 years. Often, the long maturity schedule relative to the useful life of a project works against this form of financing except in the most creditworthy municipalities.

Income on general obligation bonds is tax exempt. Section 103(a) of the Internal Revenue Code of 1954 provides that gross income does not include interest on obligations of a state or political subdivision of a state. [This same tax-exempt treatment is extended to interest on industrial development bonds used to finance solid waste disposal facilities under Section 103(b) of the Code. Pollution control revenue bonds, which are issued by a public entity on behalf of a private enterprise to enable it to obtain low-cost financing for pollution control, are also covered under the Code. They carry tax-exempt status and are similar in form to municipal revenue bonds; however, the credit rating of the corporation and its guarantee of revenues are keys to the marketability of such bonds.]

A typical general obligation bond is offered competitively for sale to bidders. A competitive bid solicitation is used to invite investment banking houses and commercial banks to make sealed bids for the right to purchase and resell the bonds. Underwriting syndicates are normally formed by groups of firms to purchase the entire issue. The bidder offering the lowest net interest rate to the municipality obtains the right to place the bonds with its customers or purchase them for its own account.

General obligation bond financing is a good option if the community is willing to take the risk of added expense if something goes wrong with the plant. It is better suited to turnkey-type procurements where risks may be mitigated somewhat through contracts with process vendors that provide certain performance guarantees and backup commitments. On the other hand, this form of financing may permit projects to be financed without all critical elements in place. Moreover, it places the primary risk on the tax-payers and is not viable for municipalities at or near their legal debt limits. The stringent limitations on local government spending today, as well as competing needs for such debt, make the issuance of general obligation debt for waste-to-energy facilities more difficult.

Municipal Revenue Bonds

The use of municipal revenue bonds in project financing shifts the security from the taxing power of the issuing entity (i.e., a municipality) to the revenues of the project-user charges (tipping fees), revenue from the sale of energy (and perhaps material) products, and interest income from reserve funds. Municipal revenue bonds are long-term obligations typically issued by a municipality or quasi-public agency such as an authority authorized to issue such debt instruments. The bonds are tax exempt under the IRS code. Typically, voter approval is not required and municipal debt limitations do not apply, since revenue bonds are not backed by the taxing power of a municipality. The issuing entity usually convenants to fix and collect rates for services provided by the project sufficient to pay operating expense, bond principal and interest, and all other payments that may be specified in the trust document. Additionally, revenue bond financing usually requires a debt service coverage of 1.1 to 1.5 times the bond principal and interest re-quirements, the creation of various reserve and contingency funds, and cer-tain insurance as additional security for the bondholders. Revenue bonds usually yield an interest rate of 50 or more basis points higher than gen-eral obligation bonds. Furthermore, they come under very close scrutiny by the investment community since they are not secured by the "full faith and credit" (i.e., taxing power) of a municipality.

Strict project financing with revenue bonds requires at the very least a "take-or-pay" energy purchase contract, a "put-or-pay" solid waste dis-posal agreement, a guarantee of the facility operating performance, provi-sion for *force majeure* events, and various types and customary levels of in-surance. Transaction costs as a percentage of total capital costs are higher than for general obligation financings. Significant additions to total capitalization are incurred as reserve funds, interest during construction, and issuance costs.

Revenue bonds are usually offered through a negotiated underwriting with one or more investment banking underwriters. They are generally issued with maturities ranging up to 30 years. In order to issue revenue bonds, the governing body of a municipality (or authority) must adopt a bond resolution specifying the application of bond proceeds to the construction of the project, creating a lien on revenues of the project, and setting forth the rights of the bondholders and the obligations of the issuer.

In some instances, municipalities have also pledged tax revenues as additional security. (This is not permitted in all states.) This is called a "double-barreled" revenue bond. While the issuing entity still relies on project revenues to pay debt service, the "full faith and credit" guarantee of the municipality allows the bonds to be considered as general obligation bonds and to sell at a lower interest rate. This arrangement is similar (although not exactly comparable) to the revenue bonds issued to finance the Harrisburg, Pennsylvania incinerator. The City created an authority who issued bonds and assumed legal title to the facility. The City then signed a noncancelable contract with the authority for bond repayment. This revenue bond backing was viewed in the investment community as analogous to the issuance of general obligation bonds by the City; thus they carried the same negotiated interest rate as G.O. bonds issued directly by the City. In many cases, however, this form of guarantee for a revenue bond project may be politically unacceptable. This approach is used to provide added security to the project in order to enhance its financeability, giving the bondholders the additional backing of the municipal taxing power.

The issuing entity, whether a municipality, municipal authority, or other authority arrangement, must have the ability to "control" the waste stream and to engage in long-term contracts with private collectors and/or other municipalities to ensure sufficient waste quantities and, thus, revenues to the project. In addition, firm, long-term contracts of a "take-or-pay" nature, whereby one or more energy markets agree to purchase a minimum quantity of energy and pay a minimum amount for it whether they use it or not, must be secured before the financing can be completed. These contracts are usually congruent with the life of the bonds.

In the case of revenue bond financing, as in other options, the project may be operated by the issuer or by a private operator under contract.

Industrial Development Revenue Bonds

Industrial development revenue bonds (IDBs) are a form of revenue bond issued through an agency acting on behalf of a municipality or group of municipalities and a private industry or other commercial entity. The issuer is usually a state-designated agency. The major distinction between industrial revenue bonds and other municipal revenue bonds is that they are not tax exempt unless they qualify for special exemptions. Under Section 103(b)(4)(E) of the Internal Revenue Code, solid waste disposal facilities (which include waste-to-energy facilities) are one of the special activities exemptions, thus, interest income on industrial revenue bonds issued to finance such projects is exempt from federal income tax. Interest income may or may not be exempt from state income taxes.

It is important to note that the U.S. Treasury regulations specify that "solid waste" to be processed, or disposed of, in a facility that is financed with tax-exempt IDBs must be "valueless" at the place it is located. Thus, if anyone is willing to purchase the solid waste at any price, it does not qualify as solid waste. Usually, in most IDB financings, an advertisement is placed in local media to ensure that no party is willing to purchase the unprocessed municipal solid waste.

Other IRS requirements issued through temporary regulations include: (1) at least 65 percent by weight or volume of the total materials introduced into any solid waste processing facility that recycles or reconstitutes them into a material that is not solid waste must be solid waste; (2) where a solid waste facility is also an energy recovery facility, the portion of the facility that performs an additional conversion process beyond the first marketable product will not qualify for tax-exempt financing (i.e., a steam pipe used to carry steam generated at the facility to a steam customer would not qualify); and (3) at least 90 percent of the net proceeds of the bond remaining after payment of issuance costs (including funded reserves, capitalized interest, and underwriter's spread) must be used to pay for those parts of the facility that qualify as "solid waste" disposal equipment. The remaining 10 percent can be used to finance any other items of the project that qualify under the specific state legislation through which the bonds are issued, including components that do not qualify as solid waste disposal facilities for federal tax purposes.

The federal Crude Oil Windfall Profit Tax Act of 1980 modifies certain tax law regarding solid waste facilities. It provides that "solid waste energy producing facilities" financed by IDBs and selling steam and electric energy to a federal agency will not lose their tax-exempt status because the financed facilities are used for the benefit of the federal government or because the debt service is derived from payments by the federal government. This is conditioned by several restrictions: (1) 90 percent or more of the fuel for the facility must be solid waste; (2) both the solid waste disposal components and the steam and electric generating components are owned and operated by the agency which issues the bonds; and (3) all steam and electricity are sold to the federal government. The agency can enter into a management agreement with a private party for operation of the facility as long as the term does not exceed one year.

Other provisions of the Crude Oil Windfall Profit Tax Act of 1980 expand the definition of solid waste disposal facilities which qualify for tax-exempt industrial development bond financing to include "qualified steam-generating facilities" of which more than half of the fuel is solid waste or fuel derived from solid waste. This has important bearing on refuse-derived fuel (RDF) facilities. To qualify as a steam-generating facility, the fuel (i.e., RDF) must be produced at a facility located at, or adjacent to, the site of the steam-generating facility, and the processing facility must be owned and operated by the party who also owns and operates the steam-generating facility. This appears to restrict the use of IDBs by an off-site industry for financing an RDF-fired steam generator (i.e., a new dedicated boiler).

In a typical project financed through IDBs, the local government(s) issuing agency technically owns the facility and the equipment, and lends the bond proceeds to a private contractor for facility construction. The actual financing structure and repayment of the bond issue are executed through a lease arrangement, an installment sale, or a loan whereby the payments are fixed in amounts sufficient to pay bond principal and interest as well as some additional amount as a safety factor (coverage). The operation and maintenance expenses are usually the obligation of the facility contractor (the lessee).

IDBs, like other revenue bonds, are supported solely by the revenues derived from the project. They are not backed by the "full faith and credit" of the issuer as are general obligation bonds. Primary revenues from the project include tipping fees from delivery of solid waste (user fees which are typically guaranteed through "put-or-pay" contracts) and monies from energy product sales. To a lesser extent, interest earnings on special funds or reserves, sales of recovered materials, and any insurance claims or liquidated damages may provide additional revenues.

The issuance of IDBs may be linked with equity from the private contractor. This can lower the amount of bonded debt needed for the project and strengthen the commitment of the private party in the project. IDB financing combined with private equity usually permits the private firm or joint venture to gain the benefits of lower-cost, tax-exempt financing as well as other tax benefits including accelerated depreciation, investment tax credit, and interest or rental deductions through constructive tax ownership. In return, the municipality(ies) can achieve a lower cost for disposal of solid waste, as the private contractor is able to pass along a portion of the tax savings in the form of a reduced tipping fee. This is a very important consideration for a municipality when evaluating public versus private ownership.

Industrial development revenue bonds are closely scrutinized by the investment community and require several security features to be marketable. The credit rating of the private industry involved in the project is very important as are the corporate guarantees. Interest rates may be over 100 basis points higher than for a comparably rated general obligation bond. Significant additions to total capitalization are incurred through reserve funds, interest during construction, underwriter's spread, and other transaction costs.

Comparisons of revenue bond sizings for a small-scale waste-to-energy project with a capital cost of $10 million and a financing period of 20 years are presented in Table 7-1. This table is provided to show the reader the impact on the annual cost of capital (and the components in the bond issue) incurred through higher interest rates and the additions to the basic facility cost (installed capital cost) inherent in the revenue bond financing. In going from a 10 percent interest rate to a 13 percent interest rate (which is more typical of the rates experienced during 1981–1982 in the bond market), the net effects are approximately $10 per input ton increase in the debt service payment and substantial additions to transaction costs. The interest rate ef-

Table 7-1. Revenue Bond Sizing—150 TPD Waste-to-Energy Project (Thousand Dollars).

	BOND INTEREST RATE			
	10%	11%	12%	13%
Total Capital Required from Bond Issue[a]	10,000	10,000	10,000	10,000
Interest Payments during Construction	2,698	3,019	3,355	3,704
One Year Debt Service Reserve[b]	1,584	1,723	1,872	2,028
Operation and Maintenance Reserve[c]	500	500	500	500
Bond Issuance Expense[d]	472	480	489	499
Less Earnings on Funds[e]	(1,765)	(1,998)	(2,236)	(2,485)
Total Bond Issue	13,489	13,724	13,980	14,246
Annual Debt Service	1,584	1,723	1,872	2,028
Per Ton[f]	36.16	39.33	42.73	46.30

[a] Represents installed capital costs of a 150 TPD modular system (1984 dollars). Construction funds are based on a drawdown schedule of four 6 month periods at 15%, 40%, 35%, and 10%.
[b] Based on a 20 year amortization period at the indicated annual interest rate.
[c] Represents operation and maintenance expense for 4 months.
[d] Represents 3½% of the total bond issue and includes underwriters fee, legal expenses, etc.
[e] Includes interest earned on a construction fund, capitalized interest fund, operation and maintenance reserve fund, and debt service reserve fund.
[f] Based on 120 TPD processed (150 TPD capacity at 80% availability) times 365 days per year resulting in an annual throughput of 43,800 tons.

fect would be similar, although much larger in actual dollar amounts, in a larger-scale project.

The point is that interest rates have a significant impact on project costs, and while the actual equipment or facility cost may be a specified amount, the total revenue bond size required to finance the project may be over one and one-half times that amount when the costs of financing are applied. It is important to recognize this when planning a waste-to-energy project.

IDBs may be used in conjunction with other private investment arrangements such as leveraged leasing, discussed in the next section. Normally, the maximum amount of bond issue permitted for IDBs is $10 million; however, under a special exemption, municipal solid waste projects have no capital expenditure limits.

The federal Tax Equity and Fiscal Responsibility Act of 1982 (TEFRA) contained significant provisions affecting waste-to-energy facilities, particularly the treatment of certain tax benefits when projects are financed with industrial development bonds. Basically, this law afforded an exemption for solid waste disposal facilities, which includes waste-to-energy systems, that permitted the full use of both industrial development bond financing and Accelerated Cost Recovery System (ACRS) depreciation. In essence, this excluded solid waste facilities from the "anti–double dipping" provisions that attend certain other types of projects financed with industrial development bonds whereby the owner(s) must use the longer-term straight line depreciation. With the longer-term depreciation, the tax benefits in early years are reduced, along with investor interest in the project.

TEFRA included other provisions that affected industrial development bond financing. A public hearing and approval by the elected officials or legislative body are now required for the issuance of industrial development bonds. This requirement includes the issuing jurisdiction and the municipality where the project is to be located. Additionally, the length of maturity on the bonds cannot exceed the weighted average estimated economic life of the assets financed by the bonds by more than 20 percent. There are other provisions that also impact projects financed by industrial development bonds.

More recently, a significant impact on financing waste-to-energy projects with industrial development bonds was created through the passage of the Deficit Reduction Act of 1984. In short, this Act limits the amount of tax-exempt industrial development bonds, commonly known as "private activity bonds," that can be issued within a particular state in any given year to no more than the greater of (1) $150 per capita multiplied by the state's population or (2) $200,000,000. The Act includes a carry-forward provision which allows issuing authorities to carry forward unused portions of their allocated share of the state's total cap amount from one year to the next for up to three years for specific projects which they intend to finance with IDBs. The state cap applies to obligations issued after December 31, 1983; however, obligations for which an inducement resolution was adopted prior to June 19, 1984 are exempted if issued on or before December 31, 1984. The cap is statutorily allocated 50 percent to state agencies and 50 percent to local issuers and, in the case of local issuers, is to be further allocated on a population basis. This allocation can be overridden by action of a state's governor, within a specified period, or by the legislature.

Additionally, the Act prohibits use of the proceeds of IDBs to finance land or existing property, requires straight line depreciation on solid waste (and sewage) projects, subjects solid waste projects to new, restrictive arbitrage rules, and includes a $40 million cap of IDBs in which one company is the beneficiary.

After December 31, 1984, arbitrage earnings during traditional temporary periods must be rebated to the United States instead of being applied to the costs of acquiring and constructing the project. This, in effect, will increase the size of bond issues and reduce the interest earnings, which will normally increase tipping fees. Exceptions to this rebate requirement are provided for bona fide debt service funds with annual gross earnings of less than $100,000 and for issues where the gross proceeds, other than the proceeds in a bona fide debt service fund, are expended within six months of the issuance date. Gross proceeds include the amount received as a result of investing the original proceeds of the issue and amounts used to pay debt service on the issue.

The restriction on depreciation or cost recovery of all property financed with the proceeds of tax-exempt bonds provides that the costs of such property must be recovered on a straight line basis over the property's Accelerated Cost Recovery System life. This applies to property placed in service after December 31, 1983 to the extent property is financed with obligations issued after October 18, 1983. Thus, in general, this provision reduces the percentage of depreciation a facility owner can claim in the earlier years of a project, forcing depreciation tax benefits to be recovered over a longer period.

It is beyond the scope of this book to provide a detailed review of tax laws. However, project developers should carefully review legislative developments with tax counsel and bond underwriters when considering the financing and ownership structure for a waste-to-energy project. The tax laws have changed many times in the last several years, and other significant tax law reform proposals are being considered at this time.

Leveraged Leasing

Leveraged leasing is a method of financing whereby tax benefits permitted by the IRS in connection with acquisition and ownership of an asset are transferred from a borrower (lessee) to a lender (lessor). Therefore, the waste-to-energy facility is owned by the lessor, not the municipality or the operator. The cost of the lease financing is based on the tax benefits of ownership transferred by the lessee to the lessor. Leveraged leasing offers another form of financing whereby the participation of private investor(s) can serve (1) to lower the amount of bonded indebtedness of the issuer and (2) to reduce the tipping fee payments of the participating municipality(ies) by passing on a portion of the tax savings realized. Leveraged leasing is typically combined with industrial development bond financing; however, other types of obligations can be used. Depending on the exact lease structure, the specific tax benefits, and the risks involved, the effective interest rate for the total financing can be reduced up to 500 basis points through leveraged leasing.[1]

This form of financing involves the participation of a third party private investor(s) who provides typically 20–30 percent of the project costs and the leveraging of that investor's funds with the tax-exempt financing of the remainder of the project costs. The tax benefits of private ownership become a significant contribution to the project. The private investor becomes the owner/lessor of the project and, in turn, leases the facilities to another private entity termed the user/lessee. In the event the user/lessee is able to take full advantage of the tax benefits of ownership, there may be little economic justification for engaging in leveraged leasing. The determination of whether or not the user/lessee can fully utilize the tax benefits requires a careful review of user/lessee's estimated tax liabilities over the period during which these tax benefits would be available.

The lease equity market is comprised of several hundred taxable institutions, including commercial banks, finance companies, and life insurance companies. The aggressiveness of these institutions in bidding to become investors in leveraged leases depends upon a number of factors, including remaining tax-shelterable income, interest in various investment categories, credit analysis perspective, and residual value expectation of various assets. In order to obtain third party equity at the lowest associated rental cost, it is essential to approach those institutions most competitive for the asset category, most receptive of the particulars of the tax benefits, and most accepting of the credit support.

In facility leases, another dimension of the equity marketing process is that considerably more negotiations and documentation would be required than are normal for such items as computers, rolling stock, and barges. For facility leases, therefore, it is helpful to solicit those equity investors who are familiar with complicated transactions. A small number of larger investors

---

1. One hundred basis points equals one percentage point.

reduces the negotiating parties, as well as increases the size of investment, which justifies the commitment of management time of the investor to the transaction.

In leveraged leasing, equity investors acquire the project to be leased prior to its in-service date for tax purposes through a specially created ownership trust or partnership (the owner trustee receives a title to the facility and becomes the lessor). Their investment must be at least 10 percent of the cost of the project. The balance of the cost of the project is obtained from the issuance of long-term obligations by a local government or other agency. The debt financing is without recourse to the equity investors and is secured only by an interest in the project assets and an assignment of the lease. The equity participants' return on investment is derived from (a) the net cash flow available from lease rentals in excess of debt service, (b) the tax benefits of ownership, and (c) the residual value of the project.

The lenders providing the debt portion of the financing look primarily to the ability of the lessee to make timely rental payments and to the collateral value of the project. The lenders have a claim on lease rentals prior to that of the equity participants. An indenture trustee collects the proceeds from the equity participants, pays for construction of the facility, and services the debt.

It is essential to a leveraged lease financing that the tax benefits of ownership accrue to the lessor. The tax benefits available to an owner of an asset in a leveraged lease usually will provide an adequate rate of return to some level of equity investment without any significant cash flow from the lessee. Since this investment replaces debt that would otherwise be provided by a bond issue, the rental payment is significantly less than debt service on 100 percent debt financing. Since interest rates on debt in a leveraged lease are usually identical to rates for direct debt financing, the rental payments are reduced in proportion to the equity investment in the lease. In a leveraged lease, the greater the present value of the tax benefits, the greater is the amount of equity provided for a required rate of return and the lower is the lease payment.

It is also important to note that security features in a lease financing, such as a guaranteed waste supply, a long-term energy customer, and strong equipment and performance guarantees by qualified vendors, are still critical to the success of this financing and will be carefully scrutinized by the investment community.

There are two significant problems with leveraged leasing. First, the equity of the lessor is usually not committed until the project has been completed and has passed performance tests. This necessitates interim financing. Second, there is a risk the equity will not be committed if design problems occur or the facility does not pass performance tests. Leveraged lease financings have been very difficult to arrange for small-scale waste-to-energy projects, and although many have been proposed, few have actually been completed. However, it is likely that we will see more leveraged lease financings for small-scale waste-to-energy projects where there is a more limited quantity of "vendor equity" available.

CREATIVE FINANCING TECHNIQUES

Over the last several years the investment banking community and waste-to-energy industry have pioneered some creative approaches to enhance project economics and ultimate financing. The techniques have been developed in response to the following:

- The escalation in bond interest rates
- The generally depressed bond market for waste-to-energy projects
- The need to minimize user charges (tipping fees) in the early years of a waste-to-energy plant's life cycle

The following brief discussion of some of these techniques is designed to familiarize the reader with the basic concepts. These techniques are typically combined with traditional bond financing methods. They generally serve to benefit all parties in the deal and make the waste-to-energy project more attractive to the community(ies) which it serves.

Deferred Equity

To establish ownership of a waste-to-energy facility, it is not necessary for a system vendor to put equity into the project during construction. Typically, a community may prefer the vendor to contribute equity during the construction phase. However, a community must recognize that this contribution is not "free." The vendor must be compensated through the project for this contribution of funds which offsets the amount that has to be raised in the bond issue. The vendor may have a required rate of return on this equity investment of 25–30 percent or more per year as compared to the bond funds which can be obtained in the current market at an interest rate of 10–12 percent. Through equity investment, the vendor expects to receive tax benefits of ownership, beneficial use of the facility, and its residual value at the end of the project. However, the vendor may prefer to put his equity into other higher-return opportunities during the term of construction of the waste-to-energy facility when tax benefits are minimal.

A creative way to obtain the leverage of 100 percent tax-exempt debt financing and still derive the tax benefits, yet share those benefits with the community, is for the vendor to "defer" equity infusion until the project is operational and then contribute equity in the form of a subsidy or rebate on the tipping fee. Under this financing approach, the capital cost of the project is usually funded entirely through the issuance of tax-exempt industrial development bonds. Once the project commences operation, the contractor rebates monthly a portion of the tipping fee. The specific rebate arrangement or formula and dollar amounts are established through agreement. This financing arrangement provides flexibility and offers incentives and risk to both contractor and community.

The contractor takes the risk that it may not have sufficient predictable tax liability in the future to use all the tax benefits under reasonable economic conditions. Further, the contractor must absorb the legal costs and time to carefully document and structure the financing using this technique. The community, or its bond issuing agency, takes the credit risk that the contractor will be able to make the rebate in the future as opposed to having the equity available during construction. The benefits, however, can mean a lower tipping fee for the community and the funding of the equity requirement out of tax-exempt debt for the contractor. Conceptually, this financing arrangement may offer a lower financing cost than leveraged leasing in view of the 100 percent leverage factor with tax-exempt debt.

Step Financing (Deferred Principal Repayment)

Another concept which has been used in financing activities other than waste-to-energy projects in the past, is the deferred repayment of the principal on the bonded debt used to finance the waste-to-energy project. In fact, this concept is normally applied during construction, as interest only is paid until facility operation when substantial cash flow to the project commences. The "step financing" involves the extension of the debt service principal payment deferral out until the fifth or later year of operation (i.e., seven to eight years after construction starts in a project), with the community paying only the interest portion of the debt service until then. The full principal repayment is then spread over a shorter period (i.e., 8–15 years versus 20 or more); however, the communities have benefited by paying a lower tipping fee in the early project years when energy sales revenues could be expected to be lower. On balance, they must absorb the debt service payout over a shorter period, and they will end up paying more interest over the life of the financing. However, the increased tipping fees, in theory, are offset by the expected higher energy sales revenues in later years. This approach to financing is being considered in several projects.

Variable Rate Financing (Floating Rate Securities)

Historically, the primary financing mechanism for waste-to-energy projects as well as other capital construction projects was through the use of long-term fixed rate debt. The debt maturities matched the economic lives of the assets financed, and interest rates were relatively low. The recent debt market has been characterized by periods of high, volatile interest rates, rapid increases in the number of issuers, and changing investor needs. These and other factors have combined to create an environment in which long-term fixed rate debt is not necessarily the best approach to capital financing, and this holds as well for waste-to-energy facility financings. As the fixed interest rates rise, waste-to-energy projects become less economically attractive. The introduction of creative debt instruments that contain a floating interest rate and are priced off the short end of the yield curve can result in substantial interest rate savings and allow certain projects to be constructed that may not be economical if financed at a higher fixed rate.

Various floating rate debt instruments have been successfully applied in the marketplace, and today there is an active market for securities that are repriced or mature from periods as short as one day to five years. Several recent waste-to-energy project financings have incorporated some form of floating rate debt. There are several specially structured securities unique to the investment banking concerns that market them; however, floating rate debt can generally be classified into variable rate demand bonds (VRDBs), adjustable rate bonds (ARBs), and daily adjustable tax-exempt securities (DATES).

Variable Rate Demand Bonds

VRDBs are long-term bonds sold at short-term interest rates. This is accomplished by giving the bondholders the right to have their bonds purchased ("put" option) at 100 percent of their principal amount or par after giving appropriate notice. Usually, the bondholder has the right to put his bonds on the first day of each month or upon seven days' notice. The bonds are purchased by a remarketing agent rather than the issuer. The remarketing agent attempts to remarket the bonds to new investors; however, if it is unable to do so, the trustee provides funds for bond purchase by drawing on a letter of credit from a financial institution. The letter of credit covers the full principal amount of the bonds, and the financial institution, normally an AAA bank, holds the bonds not remarketed until purchasers are located or the letter of credit expires, whichever is shorter. Future bond purchasers have the same right to put bonds as the original bondholders. The bondholder's put option every 30 days or on 7 days' notice allows the investor to treat the security as a short-term investment of 30 days of less. Thus, the investment banker is able to market the bonds at a short-term interest rate even though they represent long-term securities.

The interest rate is set by the remarketing agent and is based on an appropriate index such as a percentage (50–60 percent) of the prime rate published by a commercial bank, the 30 day tax-exempt commercial paper rate, or a percentage of the 91 day treasury bill rate. The interest rate is adjusted each reset period or is structured to be adjusted only when a change occurs in the index or when the securities are put. In any case, the interest rate "floats" during the term of bonds. The VRDBs can be structured to allow the issuer to convert to a fixed rate of interest upon 30 days' notice for the remaining term of the bonds. This feature would be important if market conditions were to change such that fixed rate financing is more attractive. It is important to note, however, that historically VRDBs have never exceeded the interest cost of traditional fixed rate financing.

VRDBs in the current marketplace sell at an interest cost of 6.1–6.5 percent. To this rate must be added the costs of a letter of credit. Annual fees for a letter of credit typically range up to 1 percent of the face amount of the letter of credit.

The bonds are secured by the letter of credit from a commercial bank typically rated at least AA or an insurance policy rated AAA or other ratings (i.e., MIG-1) acceptable to investors. If the primary security is a letter of credit, the project owner/operator would normally have to pledge project revenues to secure its obligation to reimburse the bank for any and all monies drawn on the letter of credit.

In view of the flexibility provided to the issuer and the investors, VRDBs have become well accepted in the tax-exempt bond market. They offer the investor security and liquidity. VRDBs are normally sold to institutional investors and pay interest on a monthly or quarterly basis. They are often privately placed to reduce the costs of issuance and the likelihood that they will be put. In the current market, a VRDB with an interest cost of 7 percent (includes letter-of-credit fee) compares very favorably to a long-term, fixed rate conventional financing at 10–11 percent interest cost.

Adjustable Rate Bonds

Similar to VRDBs, adjustable rate bonds provide floating interest rates and long-term maturities with short-term interest yields. The interest rate is adjusted less frequently, with ranges of three months to five years. Bondholders may "put" their bonds on interest rate adjustment dates. ARBs may also be converted to long-term fixed rate bonds on any adjustment date at the issuer's option. These bonds are also backed by a letter of credit. Since the interest rate adjustment period is less frequent and the investor is more restricted in the put option, ARBs trade at a higher interest rate than VRDBs.

Daily Adjustable Tax-Exempt Securities

Daily adjustable tax-exempt securities are a proprietary new form of floating rate security. Unlike other variable rate debt instruments, however, DATES are repriced daily and may be put daily by investors. This "instant liquidity" feature is attractive to investors. DATES, like VRDBs and ARBs, can be converted to a variable interest rate or fixed rate by the issuer. The fixed rate may be set for periods up to the final maturity of DATES. Variable rates are usually based on an appropriate index. If converted to fixed rate, DATES may require a new offering circular. DATES are an attractive investment to tax-exempt money market funds since they count as only a 1 day investment in calculating the average life of the fund's portfolio but provide a return tied to the 30 day tax-exempt commercial paper rate. The put feature of DATES, like other floating rate securities, is typically backed by an irrevocable bank letter of credit.

In summary, variable rate debt instruments offer a creative way to finance waste-to-energy facilities, affording perhaps a more economical arrangement than traditional long-term, fixed rate financing vehicles. The issuer pays a short-term rate for long-term debt and retains flexibility to convert to variable or fixed interest rates if market conditions change. Financing costs are also lower. The investor obtains a highly liquid security with a put option backed by a strong letter of credit.

Floating rate debt is not, however, without exposure for a development agency who may apply it as the primary financing vehicle for a waste-to-energy project. Interest rates do float and can float upwards. In the future, the market for such securities may become saturated, and the ability to remarket large volumes of bonds "put" by investors may be limited. If project revenues are pledged as security by the owner/operator against the letter of credit draw and bonds cannot be remarketed, tipping fees could rise. Nonetheless, the use of variable rate tax-exempt financing instruments should be fully explored when developing the financing plan for a waste-to-energy facility. They could mean the difference between an economically frustrated project and one that is economically attractive to participating communities when translated into an interest cost savings reflected in tipping fee reductions of perhaps $10 per input ton or more.

FEDERAL TAX BENEFITS
OF PRIVATE OWNERSHIP

The following discussion provides a general description of federal income tax benefits as they relate to the private ownership of waste-to-energy facilities. Under certain conditions relating to the tax law, private ownership in a waste-to-energy facility could have a favorable impact on ultimate project cost to participating communities and shift the risks of ownership to the private party. However, relevant tax law is very complex, and any misstructuring of the project could result in significant reduction or elimination of perceived tax benefits.

In waste-to-energy project financings involving private equity participants, regardless of whether the financing is structured as a loan, an installment sale, or a lease, for federal income tax purposes the transaction is viewed as a loan of the proceeds by the bond issuer to the private company. The private company would therefore be deemed the "tax owner" (or "beneficial' or "constructive owner") of the facility and would be entitled to a number of federal tax benefits. The preservation of these benefits is the primary tax objective of the private company, and care must be exercised in structuring the agreements with the bond issuer to insure their preservation.

Ownership for tax purposes is not solely or necessarily dependent upon legal title to the property but upon an equity investment in the property. In making a determination of tax owner, the IRS will consider legal title, capital investment, responsibility for operating and maintenance costs, responsibility for tax payments, responsibility for natural catastrophies and operational problems, and entitlement to project benefits. When a private operator or equity participant is identified as the "tax owner," a number of federal benefits may become available. Figure 7-1 depicts the tax benefits and their general impact on costs of a project. In general, the Tax Equity and Fiscal Responsibility Act of 1982 (TEFRA) diminished tax benefits of private ownership for waste-to-energy projects. It froze accelerated depreciation at the original 150 percent declining balance basis, eliminating the 200 percent declining balance depreciation that was to have been permitted after 1985. Cost recovery periods for property placed in service after 1983 financed with tax-exempt debt have been changed by the Deficit Reduction Act of 1984. Such property must now be depreciated on a straight line basis over its Accelerated Cost Recovery period. Both changes tend to alter the effect of depreciation benefits in the future and over the long term, reducing the net present value of those benefits. TEFRA also reduced allowed depreciation by one-half the amount of the tax credits claimed (alternatively, the taxpayer could reduce his or her ITC basis by two points). Previously, the amount of depreciation that could be expensed was unaffected by tax credits. Further,

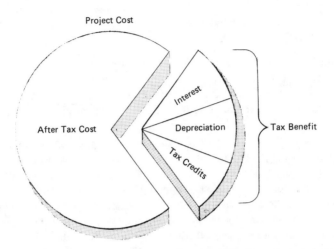

Note: Cost-Benefit proportions vary for each project.

Figure 7-1. Tax benefits of private ownership.

TEFRA requires the amortization of interest and tax expenses during construction over a ten year period. Previously, these expenses could be deducted from income immediately.

The Deficit Reduction Act of 1984 includes provisions which distinguish service contracts from leases for federal income tax purposes. The classification of a contract for the operation of a waste-to-energy facility and a contract for the sale of power from the facility is critical to the availability of tax benefits to the owner of the facility. Treatment of a service contract as a lease by the IRS could result in loss of tax benefits to the owner (either the service provider or a third party owner/lessor that leases the facility to a service provider). This potential problem is amplified if the party receiving the service is a tax-exempt entity or a public utility.

Before the passage of the Deficit Reduction Act, the investment tax credit was denied for property used by, or leased to, a tax-exempt entity such as a municipality under the "nontaxable use restriction." This presented certain problems in structuring waste disposal service contracts. If the owner of the facility were treated by the IRS as having leased the facility to a municipality rather than as having provided a waste disposal service, the investment tax credit (and the energy tax credit, as applicable) was denied and receipts to the owner were considered as rental income.

In the case in which a service is provided to a public utility, a similar problem arises. Public utility property is not eligible for energy tax credits or five year ACRS deductions. If a power sales agreement were classified as a lease by the IRS, the waste-to-energy facility could be treated as public utility property owned by a nonutility taxpayer, with a concomitant loss of the energy tax credit and certain ACRS deductions.

Under the Deficit Reduction Act, two rules are provided: the General Service Contract Rule and the special rule (Wallop Amendment) which are used to distinguish between a service contract and a lease. The Wallop Amendment, not the General Service Contract Rule, is used to determine the service contract versus lease classification for qualified solid waste disposal, cogeneration, alternative energy, and water treatment works facilities. A "qualified solid waste disposal facility" is one that provides solid waste disposal services for residents of one or more governmental units if substantially all of the waste processed at such facility is collected from the general public (including any commercial business if the waste collected from such business is collected in the business capacity as a member of the general public). If a service contract does not meet the requirements of the Wallop Amendment, it will probably not qualify as a service contract under the General Service Contract Rule. A contract involving one of the above-mentioned types of facilities will be treated as a bona fide service contract and not as a lease under the Wallop Amendment if it meets all of the four following statutory requirements:

1. Neither the service recipient nor a related entity can operate the facility. This does not prevent the service recipient from inspecting the facility, taking action in the event of a breach of contract by the service provider, or exercising sovereign power. For example, the service recipient could operate the facility until a new operator is found under an emergency situation. The length of time it could continue to operate the facility and the extent of its authority in selecting a new owner are uncertain.

2. Neither the service recipient nor a related entity can bear a significant financial burden if there is nonperformance under the service contract (other than for reasons beyond the control of the service provider). (This does not apply to temporary shutdowns for repairs or improvements; however, it does indicate that a strict "hell or high water" contract would violate the requirement.)

3. Neither the service recipient nor a related entity can receive any significant financial benefit if the operating costs of the facility are less than those contemplated in the service contract. (This does not apply to increase in financial benefits to the service recipient occasioned by increase in facility efficiency, recovery of energy or other products, or a change in law which reduces its payments. Under the Wallop Amendment, contracts in which the service recipient pays all or a portion of a

facility's "actual" operating and maintenance cost as a pass-through will not qualify. However, contracts whereby the service recipient pays a fixed operating and maintenance cost subject to escalation would be acceptable. Moreover, contracts whereby the service recipient receives a share of the "net" profits would not qualify, but contracts providing for sharing of "gross" profits would qualify.)

4. Neither the service recipient nor a related entity can have an option or an obligation to purchase all or part of the facility at a price other than fair market value. (Fair market value is determined according to a formula which the parties to the contract expect to yield a fair market value when an option to purchase is exercised, or it is determined at the time a purchase option is exercised. In a leveraged or single investor lease transaction, IRS rules for "true leases" require the purchase option to be the fair market value as of the date the option is exercised.)

Under the General Service Contract Rule, six factors are applied to determine whether an arrangement is a lease or a service contract. The factors include the following: (1) whether the service recipient is in physical possession of the property, (2) whether the service recipient controls the property, (3) whether the service recipient has a significant economic or possessory interest in the property, (4) whether the service provider bears any significant economic risk for nonperformance under the contract, (5) whether there is concurrent use of the property by the service provider to provide services to others and (6) whether the total contract price under service contract substantially exceeds the rental value of the property.

If a service contract is characterized as a lease, the type of service recipient affected will have a bearing on the extent of tax benefits lost or reduced. For example, if the service recipient is a tax-exempt entity, the owner of the facility would be denied investment and energy tax credits on facility equipment, and that equipment would have to be depreciated over its asset depreciation range (ADR) midpoint life (this is 12 years if the property has no ADR midpoint life) or 125 percent of the lease term, whichever is greater. If the service recipient is a public utility, the property would be treated as used by the public utility, and this would result in reduced ACRS deductions and loss of ETC (and, possibly, loss of ITC as well). If the service recipient is a taxable entity, the facility owner's rents could be classified as personal holding company income, and if the owner was not a corporation, it would be subject to noncorporate lessor rules. These rules could cause a loss of proportionate share of ITC and ETC for individuals and Subchapter S corporations.

A more detailed description of the tax benefits of private ownership follows.

ACCELERATED COST RECOVERY SYSTEM DEPRECIATION

The Economic Recovery Tax Act of 1981 (ERTA) provided an Accelerated Cost Recovery System (ACRS) to replace depreciation deductions permitted under prior law. It is applicable to property acquired after December 31, 1980. It established recovery periods of 3, 5, 10, and 15 years versus the myriad of longer recovery periods under the older Asset Depreciation Range of the prior tax law. Most vehicles and short-lived machinery and equipment qualify for 3 year depreciation. Generally, all other equipment, with a few exceptions, qualifies for 5 year depreciation, and real property (i.e., buildings/structures) can be depreciated over 15 years. The annual deduction is determined by applying specified recovery percentages to the cost basis of the property, which includes engineering and design costs and capitalized interest on the debt financing. Tables 7-2 and 7-3 display the recovery percentages applicable to equipment (personal property) and real property. Note: In general, the Deficit Reduction Act of 1984 requires that property financed with tax-exempt debt must be depreciated on a straight

Table 7-2. Accelerated Cost Recovery System Recovery Percentage Rates.[a]

	THE APPLICABLE PERCENTAGE FOR THE CLASS OF PROPERTY IS:			
IF THE RECOVERY YEAR IS	3 YEAR	5 YEAR	10 YEAR	15 YEAR PUBLIC UTILITY PROPERTY
1	25	15	8	5
2	38	22	14	10
3	37	21	12	9
4		21	10	8
5		21	10	7
6			10	7
7			9	6
8			9	6
9			9	6
10			9	6
11				6
12				6
13				6
14				6
15				6

[a]Source: Deloitte, Haskins & Sells.

Table 7-3. All Real Estate (Except Low-Income Housing).[a]

THE APPLICATION PERCENTAGE IS:

(Use the column for the month in the first year the property is placed in service)

IF THE RECOVERY YEAR IS:	1	2	3	4	5	6	7	8	9	10	11	12
1	12	11	10	9	8	7	6	5	4	3	2	1
2	10	10	11	11	11	11	11	11	11	11	11	12
3	9	9	9	9	10	10	10	10	10	10	10	10
4	8	8	8	8	8	8	9	9	9	9	9	9
5	7	7	7	7	7	7	8	8	8	8	8	8
6	6	6	6	6	7	7	7	7	7	7	7	7
7	6	6	6	6	6	6	6	6	6	6	6	6
8	6	6	6	6	6	6	5	6	6	6	6	6
9	6	6	6	6	5	6	5	5	5	6	6	6
10	5	6	5	6	5	5	5	5	5	5	6	5
11	5	5	5	5	5	5	5	5	5	5	5	5
12	5	5	5	5	5	5	5	5	5	5	5	5
13	5	5	5	5	5	5	5	5	5	5	5	5
14	5	5	5	5	5	5	5	5	5	5	5	5
15	5	5	5	5	5	5	5	5	5	5	5	5
16	—	—	1	1	2	2	3	3	4	4	4	5

[a]Source: Deloitte, Haskins & Sells.

line basis over its ACRS life, as opposed to the more rapid set percentage write-offs reflected in Tables 7-2 and 7-3.

Land is not a depreciable asset, and structural components of buildings are treated as real property unless they are so closely related to equipment housed that they can be considered as equipment. ACRS deductions start the year property is placed in service (ready and available for intended use). The ACRS provisions of ERTA were included in the Tax Equity and Fiscal Responsibility Act of 1982; however, as noted, the basis of depreciation is reduced by one-half the amount of the tax credit claimed. Typically, 80 to 90 percent of a project's costs qualify for five year depreciation under ACRS. However, under an aggressive tax posture, as much as 95 percent of a project's costs is being claimed to qualify by some waste-to-energy system vendors in projects in which they are the "tax owner." The prospect of recapture of prior tax benefits claimed by tax owners demands caution in structuring project financing.

Investment Tax Credits

The investment tax credit (ITC) consists of two elements: the regular tax credit and the energy tax credit.

Regular Tax Credit.   The regular investment tax credit is 10 percent of the cost of qualifying property (Section 38 property) as a direct deduction from income taxes. The credit may be used to offset the first $25,000 of tax liability plus 85 percent of the liability in excess of $25,000 beginning on January 1, 1983 due to the Tax Equity and Fiscal Responsibility Act. Qualifying property includes tangible personal property (IRS Code Section 38 property) such as equipment. It does not include buildings, structural components, or other real property. Typically, 60–75 percent of total project costs would qualify for investment tax credits. However, as noted, some vendors, on opinion of tax counsel, are claiming that costs of qualifying equipment constitute over 90 percent of a project's costs; thus the ITC could be claimed on this larger percentage if the IRS agrees with the tax filing.

The ITC is taken in the year the property is placed in service. However, for property with an estimated schedule of construction over two years, credit may be taken as construction expenditures are made on a pro rata basis. It is significant to note that the ITC is not allowed if property is owned, used, or leased by a state or local government or a tax-exempt organization. By strict interpretation, property used by a political entity or the government is not considered IRS Code Section 38 property. In specific court rulings, the IRS has confined this issue to a determination of whether the agreement between a private owner and a political entity is a lease or a service agreement. That is to distinguish whether the owner uses the property to provide a service to itself. In any event, it is critical to structure the arrangement such that a municipality or government entity is not the owner or operator of the project or does not control its use and operation or share financial risk if the investment tax credit is to be clearly preserved. This is usually done by structuring "arms-length service contracts." However, it is most important to note that the 1984 Deficit Reduction Act placed severe restrictions on the application of such "service contracts."

Energy Tax Credit.   A special energy tax credit (ETC) of 10 percent of the cost of qualified property is provided under current tax law. This credit is reduced to 5 percent to the extent of the portion of the cost financed with tax-exempt bonds (this is the anit–double dipping provision in the tax law which forces the investor to choose between the full tax credit and the tax-exempt financing or government subsidy). After December 31, 1982, no ETC was available to the extent a facility was financed with tax-exempt bonds or other government subsidized financing. After 1982, for purposes of determining the energy tax credit, the qualified investment in property was reduced by a given fraction. The fraction equals one minus another fraction—the numerator of which is that portion of the qualified investment financed by tax exempt IDBs or a governmental program, and the denominator of which is the total qualified investment. Therefore, if energy property is totally financed by tax exempt IDBs or subsidized financing, the energy credit is zero.

Subsidized financing reduces dollar for dollar the qualified investment in energy property; no portion of it is allocated to non–energy property. In contrast, IDB financing is ratably allocated to both energy and non–energy property as there is no restriction on the use of such proceeds. Subsidized financing does not include loan or price guarantees, purchase commitments, state or local tax credits, nontaxable government grants, and taxable energy grants. While the ETC was not generally available after 1982, it continued for "biomass" property through December 31, 1985—and the Business Energy Tax Credit (BETC) through 1988—and for certain long-term (at least two years) construction projects, for most energy property other than biomass property, through December 31, 1990. These longer-term projects must meet two requirements:

1. All engineering studies and permit applications must be made before 1983 (1988 for the BETC).
2. Prior to 1986 (1988 for the BETC), there must be binding contracts for acquisition or construction of project property equaling at least 50 percent of total equipment costs.

Generally, equipment such as boilers, burners, and equipment used to sort and prepare solid waste for recycling, including conversion of solid waste into a fuel or useful energy such as steam or electricity, qualifies. This does not include equipment used in the process after the first marketable product is produced. For biomass property, any alternative energy property that converts biomass (any organic substance other than oil, natural gas, or coal) to a synthetic solid fuel or alcohol for fuel purposes, as well as a boiler or burner which uses biomass as a primary fuel, qualifies. Since solid waste is a form of biomass, there is some overlap, and although it may be classified under several categories, it is only eligible for the ETC one time. Other conditions relating to the regular tax credit, such as recapture, timing, and user-limitations, also apply to the ETC.

Interest Deductions

Depending on how the various agreements among project parties are structured, the taxpayer may, subject to IRS regulations, normally deduct the interest portion of the lease payments (i.e., interest portion of the debt service) from book income or may charge the full rental payment to expense in the period incurred. Thus, additional annual tax deductions are generated through interest expense or rental charges.

In summary, it is important not only to recognize what tax benefits are available but to structure any financing arrangement, including all agreements, so that the benefits are preserved and maximized by the private party(ies) involved in the project and a portion of the benefits can be passed through to the municipalities being served by the project in the form of lower costs of service.

TYPICAL STEPS NECESSARY TO BRING A BOND ISSUE TO MARKET

It is interesting to note the typical steps needed to bring a bond issue to market. This is a process requiring critical timing and careful orchestration of all parties involved. Table 7–4 shows the sequencing and activity of each

Table 7–4. The Steps Necessary to Bring a Bond Issue to Market.[a]

ITEM NUMBER	
1.	*Preparation and Review of Bond Resolution/Financing Agreements*
	Prepared by Bond Counsel and General Counsel and approved by Issuer, Underwriter, and Underwriter's Counsel.
2.	*Establishment of Tentative Terms of Issue*
	Determined by Underwriter's Counsel and reviewed by Issuer, General Counsel, Underwriter, and Bond Counsel.
3.	*Preparation and Review of Official Statement*
	Prepared by Underwriter's Counsel and reviewed by Issuer, General Counsel, Underwriter, and Bond Counsel.
4.	*Printing of Bond Resolution and Preliminary Official Statement*
	Ordered after completion of Review.
6.	*Preparation and Review of Bond Purchase Contract*
	Preparation by Underwriter's Counsel and reviewed by Bond Counsel, Issuer, General Counsel, and Managing Underwriter.
7.	*Preparation and Review of Agreement among Underwriters*
	Prepared by Underwriter's Counsel and reviewed by Managing Underwriter.
8.	*Blue Sky Survey (State Securities Laws Requirements) and Legal Investment Memorandum*
	Usually handled by a law firm specializing in Blue Sky Registration, such action required by state statutes to permit the marketing of the securities.
9.	*Invitation to Prospective Underwriters and Distribution of Documents*
	Sent by Managing Underwriter after consultation with, and approval by, Issuer. Documents distributed include Preliminary Official Statement and proposed Underwriter's documents.

(cont.)

Table 7–4. (continued)

ITEM NUMBER	
10.	*Distribution of Preliminary Official Statement to Institutional Investors and Syndicate*
	Handled by Managing Underwriter.
11.	*Information Meetings with Underwriters*
	Held in those cities where there is a sufficient number of potential underwriters. The Managing Underwriter, Issuer, Bond Counsel, and Underwriter's Counsel attend these meetings.
12.	*Information Meetings with Potential Institutional Buyers*
	Held in those cities where there is a sufficient number of potential institutional buyers. The Managing Underwriter, Issuer, Bond Counsel, and Underwriter's Counsel attend these meetings.
13.	*Tentative Pricing of Issue and Solicitation of Pre-Sale Institutional Interest and Solicitation of Orders for Trusteeship—Depository and Paying Agency*
	Handled by the Managing Underwriter, but with additional solicitations by the other members of the Underwriting Account. In all stages, the Issuer is kept advised of price thinking, present indications of institutional interests, market conditions, etc.
14.	*Release of Final Pricing and Terms to the Underwriting Account Members*
	After consultation with, and preliminary approval by, Issuer, handled by the Managing Underwriter at a meeting of the Underwriting Accounting, with verbal acceptance or rejection by the members of the Account.
15.	*Receipt of Acceptance from Underwriters*
	Signed agreement among Underwriters received from each member of the Underwriting Group by the Managing Underwriter, stating that they will participate in the underwriting.
16.	*Submission of Proposal to Purchase to the Issuer*
	Handled by the Managing Underwriter in person at a meeting of the Board of the issuing agency. Meeting usually takes place in the morning (10:00–11:00 A.M.) with the immediate acceptance or rejection of the proposal by the Issuer.
17.	*Release by Telegram to the Underwriting Account*
	Sent by the Managing Underwriter to the members of the Underwriting Group, confirming final pricing, coupon rate, underwriters' spread, management fees, and other terms of the offering.
18.	*Bond Printing Ordered*
19.	*Printing of Final Official Statement*
	Distributed by Managing Underwriter to members of Underwriting Account, institutional investors, Issuer, etc.
20.	*Allotments and Confirmations of Orders to Members and Customers*
	Determined and distributed by Managing Underwriter after the proposal is accepted. Allotments and confirmations of orders are received by members at the opening of business the morning following the sale date.

Table 7–4. (continued)

ITEM NUMBER	
21.	*Signing of Bonds and Dry-Run Closing*
	Signed by one official of the Issuer. The same day all documents are checked by the Bond Counsel for the Underwriters to be certain everything is in order.
22.	*Delivery of, and Payment for, the Bonds*
	Signing of the closing documents. The Issuer is paid the purchase price plus accrued interest, and the bonds are delivered to the Managing Underwriter for re-delivery to the members of the Underwriting Account and the investors.
23.	*Preparation of List of Group Purchases and Sales by Members of the Underwriting Account*
	Prepared by Managing Underwriter and distributed to the members of the Underwriting Account and to the Issuer.
24.	*Preparation of Syndicate Profit Breakdown*
	Prepared by Managing Underwriter and sent to Issuer.
25.	*Managing Underwriter Follow-through*
	Establishment and monitoring of secondary market trading. Continuing information and operating results of Issuer are provided to rating agencies, dealers, institutional investors, etc.

[a]Source: Blyth Eastman Paine Webber, Inc. and Pryor, Cashman, Sherman & Flynn, *Resource Recovery Procurement and Financing,* New York State Department of Environmental Conservation, Albany, NY, September 1980.

step in the process. The table is excerpted from *Resource Recovery Procurement and Financing,* New York State Department of Environmental Conservation, September 1980. This document was prepared by the Department of Environmental Conservation with the assistance of Blyth Eastman Paine Webber, Inc. and Pryor, Cashman, Sherman & Flynn. It is provided as further information for the reader in understanding the many facets of waste-to-energy project financing.

CRITICAL PROJECT ELEMENTS EVALUATED BY THE INVESTMENT BANKING COMMUNITY TO DETERMINE THE FINANCEABILITY OF A WASTE-TO-ENERGY PROJECT[2]

Investors in resource recovery are usually not equity risk takers. They are typically large commercial banks, insurance companies, or wealthy in-

2. Part of this discussion has been excerpted from the report entitled *Financing Alternatives for the City of Berkeley's Proposed Resource Recovery Project,* Blyth Eastman Paine Webber, May

dividuals seeking fixed income in a secure undertaking. They want assurance of repayment by the project sponsor (a community or development agency), another party such as a private developer, and/or the operator or the project itself. There are several elements which must be satisfied in a project financed by revenue bonds in order for it to be attractive to the investment community. It is helpful to summarize those salient requirements which a bondholder seeks and which must be considered in structuring the financing for a project. In this way, community officials, development agencies, project planners, and other parties involved in a project can understand the key security features the investor looks for and can consider these elements as they plan and evaluate financing alternatives for waste-to-energy projects. These elements are presented in the following sections.

Credit Backing

Strong credit backing in the form of guarantees from a system contractor, from a project participant such as an energy user, or from the municipality(ies) involved in the project must be provided. Such guarantees can be contracts obligating project participants to perform functions to satisfy cash flow requirements. In a turnkey or full-service procurement, a contractor must commit to firm guarantees for timely construction, operating performance, and operating costs and must be willing to assume liability for failure to meet guarantees. Thus, a contractor (constructor/operator) must demonstrate corporate financial capacity either through itself, a parent firm, a partnership arrangement, or a combination of these. In the case of a general obligation financing, the taxing power of the municipal jurisdiction(s) serves as the credit backing. Bond rating agencies will carefully assess the credit backing, whether public or private, on any bond issue.

Demonstrated Expertise of Project Contractors and/or Engineers

The expertise of the engineers designing the project and the contractors building it must be well demonstrated to the investment community and the bond underwriters. There must be assurance that the project will be properly operated and maintained, and generally the contractor (if also the operator) must offer proven technology and a proven operations track record. Addi-

---

1980. It is supplemented by information obtained from various discussions with investment bankers, rating agencies, insurance companies, and others involved in the financing of waste-to-energy projects.

tionally, the project contractor(s) should be financially strong or have sufficient financial backing to support all necessary guarantees and contractual obligations. This is especially important in evaluating the quality of guarantees provided by a system contractor. There are many vendors in the marketplace, and one must assess their track records and financial stability.

Independent Feasibility Study

In a revenue bond financing there must be an independent (i.e., third party) feasibility assessment indicating favorable economics and technical viability. The study must measure life cycle costs on an aggregate basis and show that cost of disposal is reasonable compared to other alternatives or else show that there are no alternatives. Other key elements of the study include a detailed assessment of the waste supply, a contract analysis, an environmental and planning analysis, and a thorough evaluation of the combustion technology and steam/power system. This study is usually termed the Consulting Engineer's Report for the Official Bond Statement. There has been a recent trend to have this report prepared by consultants who are advising the development agency throughout the procurement and who may not be totally independent of the project. Thus, the term "independent third party" may not apply to all projects.

Backup Landfill or Other Facility Capacity

There must be available a suitable landfill or standby facility to accept residue and to provide for bypassed raw refuse disposal in the event of a facility outage. Alternative energy sources must be available if the project's energy sales contract requires "guaranteed" delivery of energy. Too, since a waste-to-energy project cannot process all solid waste generated in a community, there will always be a need to haul certain wastes directly to landfill. Therefore, a landfill is an essential part of the "project" (and a community's solid waste management system).

Revenue Sufficiency

Project revenues from tipping fees, energy sales, and other sources (i.e., taxes other than real property taxes, interest income, etc.) must be sufficient to pay for the operation and maintenance of the facility, service and debt, and provide reasonable coverage (typically 1.10–1.50 times annual debt service after operating and maintenance costs).

Reserves

In revenue bond financing, reserves for debt service (usually one year's maximum payment) and perhaps for operation and maintenance, equipment replacement, or unplanned facility modification are required. Depending on the risks in the project and the other forms of security provided, reserve requirements could be substantial. (For example, in a $10 million financing, debt service reserves, operations and maintenance (O&M) reserves and other reserves could exceed $2 million.)

Adequate Sources of Capital

Sufficient capital must be available to insure facility completion including start-up, testing, and achieving performance as specified. Such sources could be combinations of public and private debt, grants, or equity contributions. Sufficient capital is critical to meet necessary interest payments and other expenses in the event of construction delays or difficulties and delays in the early start-up period when problems could be anticipated.

Continuous Supply of Solid Waste

The project must be guaranteed sufficient feedstock and disposal fees over its life of indebtedness. Typically, the municipal jurisdiction(s) must contractually convenant to "put-or-pay" the required tonnage at variable tipping fees sufficient to cover increased operating costs or loss of revenues. The availability of a sufficient supply of waste during facility testing prior to full commercial operation must also be anticipated.

Long-Term Market(s) for Energy

A long-term contract, usually on a "take-or-pay" basis, must be secured for the purchase of energy from the project over its period of indebtedness. Typically, the contract price must be consistent with financial projections, and it should obligate the purchaser to take a minimum quantity of energy on an annual basis and pay for it even though the purchaser may not be in a position to use it when it is available.

Insurance

Adequate insurance must be maintained during construction and operation of the project. The basic program must provide relief for facility outage and business interruption as well as liability coverage.

As the resource recovery industry has evolved, so too have the insurance industry and the types and levels of insurance that it will now underwrite for waste-to-energy facilities. This evolution has stemmed largely from the investor demands for more security and the ability of the industry to better identify, quantify, and mitigate the risks in a waste-to-energy project. The application of "municipal bond insurance" and "system performance insurance," or "efficacy insurance" as it is commonly referred to in the trade, to project financings has served to enhance, if not enable, the financing of recent projects at acceptable interest rates. Insurance carriers see an expanding market for such coverage as the financing for waste-to-energy projects trends away from general obligation debt in favor of revenue bond financing.

Insurance offers the ability to mitigate risks in a waste-to-energy project. While insurance is important in protecting the economic interests of the project owner/operator, it also has a role in satisfying and protecting the interests of the bondholder and the energy purchaser. In fact, in waste-to-energy projects the continuation of service is extremely important. It is useful to discuss briefly some trends in insurance for waste-to-energy projects since there has been considerable interest recently in certain forms of insurance as a method of "security enhancement."

Municipal bond insurance is being used more frequently in waste-to-energy plant financings, although it has been applied to other municipal and utility-type bonds for years. It is provided in projects to insure the payment of principal and interest on the bonds in the event that revenues and reserves are insufficient to make a required payment on any future scheduled payment dates.

Municipal bond insurance extends over the term of the bonds and, once issued, is noncancelable. The premium is a one-time payment up front which is usually capitalized as part of the bond issue. Municipal bond insurance can elevate the rating of project revenue bonds and, thus, lower the bond interest rate. The trade-off is the high up-front premium.

System performance insurance, termed "efficacy" insurance, is a form of insurance that covers the risk of loss of revenues from plant outage associated with the failure of equipment to perform as expected and produce the levels of salable steam and/or electricity as projected. The coverage requires an independent engineering assessment of the reliability of the plant systems to perform as expected. Currently, there are several carriers who offer this type of insurance, and they reinsure through several of the large multiple line insurance companies.

The insurance covers design performance guaranteeing O&M and debt service payments as well as the cost to modify the project to bring it up to satisfactory levels of performance. It also covers errors in design. All risk coverage with construction delay endorsement and boiler and machinery insurance must be in place before efficacy insurance will be provided.

The insurance usually does not cover 100 percent of the design capacity. Moreover, the carriers typically require co-insurance, a reserve fund and/or letter(s) of credit in the amount of annual debt service as a deductible, and retention and performance guarantees by the equipment suppliers (i.e., boiler/turbine manufacturers), which would first be applied to a loss before efficacy insurance payments are infused. Carriers of this insurance also require an advance fully paid premium.

As noted, other forms of insurance including builder's risk, construction delay, boiler and machinery, workers' compensation, general liability, and business interruption are also applied in waste-to-energy facilities. It is important to remember that insurance does not replace strong performance guarantees, sound technology, and good project management in a waste-to-energy project.

Demonstrated Compliance with Laws and Regulations

The facility as planned, designed, and constructed must meet all applicable federal, state, and local laws and regulations. This is particularly important in view of the track record of several past projects where costly modifications had to be made in order for the facilities to meet pollution control laws, particularly as they relate to emissions to the air. The investment community has become very sensitive to this element in recent years.

Favorable Rating by Recognized Rating Agencies

In order for the bonds to be marketable they must receive at least an "investment grade" rating by one or more recognized bond rating agencies. Ratings are made on the basis of financial strength of the guarantor of the debt or the perceived ability of the revenues to meet debt payment. Most revenue bonds used to finance waste-to-energy projects have been rated by either or both of the two major rating agencies: Moody's Investors Services or Standard and Poor's Corporation.

8
ENVIRONMENTAL ISSUES

The use of municipal solid waste (MSW) as fuel or feedstock in energy recovery systems can be viewed from two opposing environmental perspectives. One view is that the quantity of refuse for disposal is diminished and inactivated to a great degree, thus reducing negative environmental effects. In addition, recovery of energy and materials provides added benefit by substituting for virgin raw materials. However, the recovery of energy from waste has a negative effect common to most industrial processes—that of producing residues which require control and disposal.

In this chapter, the environmental impacts of energy recovery are addressed (what types of air, water, and solid pollutants are associated with energy recovery systems). Applicable federal and selected state environmental regulations are reviewed. Finally, technologies used to control such pollutants are described.

SOURCES OF AIR POLLUTANTS

Combustion of solid waste or refuse-derived fuel (RDF) is the major cause of air pollution associated with energy recovery systems. The characteristics of flue gas, before entering the air pollution control device, depend on various factors, including the following:

- Composition of refuse or RDF feed
- Combustion temperature
- Amount and flow patterns of excess combustion air
- Combustion approach: grate (mass burning), spreader stoker (RDF dedicated boiler), or suspension (RDF co-fire)
- Flue gas velocities
- Combustion residence time

In addition to stack gas, there are air emissions due to vehicle traffic, i.e., trucks that deliver refuse or RDF to a facility and haul residue to a disposal site. The amount of such emissions is very site specific and typically involves trade-offs with the existing waste hauling system. RDF production facilities also emit certain air pollutants (e.g., fugitive dusts), but such emissions tend to be minor.

TYPES OF AIR POLLUTANTS

A variety of pollutants is emitted as a result of the solid waste combustion process. Most are produced at a very small rate. Eleven pollutants are reviewed in this chapter:

CRITERIA POLLUTANTS	OTHER
Particulates	Hydrogen Chloride (HCl)
Nitrogen Oxides (NO_x)	Hydrogen Fluoride (HF)
Sulfur Oxides (SO_x)	Beryllium (Be)
Carbon Monoxide (CO)	Mercury (Hg)
Total Hydrocarbons (THC)	Dioxins
Lead (Pb)	

The pollutants in the first column, together with ozone, form a group for which national ambient air quality standards have been established (see below for discussion).

Particulates

Particulate levels are related to degree of turbulence in the combustion chamber. Modular incinerators, which operate at less than stoichiometric conditions in the primary chamber, typically have low particulate levels. Waterwall incinerators for either direct combustion of refuse or RDF have lower particulate levels than commonly associated with traditional refractory-lined incinerators. The large volume of combustion air required to cool refractory-lined incinerator walls creates a high degree of turbulence in the furnace. In any case, pollution control equipment exists that can control particulate emissions.

Nitric Oxides (NO_x)

Emissions of NO_x from combustion sources are the result of either the conversion of nitrogen in the fuel to nitric oxides or the fixation of atmospheric nitrogen at high temperatures. It is believed that nitric oxide (NO) is formed mainly on the flame front where temperature is high and oxygen available. The NO formed in the furnace subsequently oxidizes to nitrogen dioxide (NO_2) in the atmosphere. Generally, production of NO_x from burning refuse is low due to low operating temperatures.

Sulfur Oxide (SO_x)

Emissions of SO_x upon incineration of refuse are a function of sulfur content. Typically, sulfur levels in solid waste are low, about 0.2 percent on as-received basis.[1] By comparison, the average sulfur content of western coal is 0.4 percent and eastern coal is in excess of 1 percent.

Carbon Monoxide (CO)

Carbon monoxide is the most abundant gaseous pollutant emitted upon incineration of municipal solid waste; it is a product of incomplete combustion. Improved design of overfire nozzles in the combustion chamber of direct combustion and RDF furnaces has reduced the amount of incomplete combustion.

Total Hydrocarbons (THC)

Another product of incomplete combustion is THC, usually in the form of low-molecular-weight hydrocarbons, aldehydes, and organic acids. This pollutant combines with nitrogen oxides to form photochemical oxidants, or smog, under warm, sunny conditions. Hydrocarbons and NO_x are precursors to ozone, a prominent constituent of smog. Since ozone is highly toxic, efforts are directed to control THC and NO_x.

Lead (Pb)

Some of the lead in municipal waste is emitted with fly ash from incineration. The amount of lead will depend on the efficiency of the particulate control system.

Hydrogen Chloride (HCl) and Hydrogen Fluoride (HF)

HCl and HF are by-products of the combustion of chlorinated and fluorinated plastics. The primary source of chlorides is polyvinyl chloride. Since the percentage of plastics in the waste stream is forecast to grow, increased emission of these two pollutants is possible. Growth in plastic discards and related air emissions from combustion may be offset if recycling of plastics becomes common practice.

---

1. E.R. Kaiser, "Physical-Chemical Properties of Municipal Refuse," in *Proceedings of the 1975 Louisville Symposium on Energy Recovery from Refuse,* University of Louisville, 1975.

Beryllium (Be) and Mercury (Hg)

Beryllium and Mercury are heavy metals and can be considered toxic pollutants. Emission standards for mercury have been set by the U.S. Environmental Protection Agency (EPA) under the National Emission Standards for Hazardous Air Pollutants (40 CFR 61.52).

Dioxins

Emission of dioxins from an RDF-fired boiler in Hempstead, New York was partly responsible for closing the facility. Existing research indicates that there is significant uncertainty regarding the degree of hazard posed by dioxin emissions. Dioxin is a generic term for a family of organic compounds. Polychlorinated dibenzo-*p*-dioxins (PCDD), especially tetra-CDD (TCDD), have received most attention from an environmental health perspective. One isomer of TCDD (2,3,7,8) has been reported very toxic. This is the type of dioxin contained in Agent Orange, a defoliant used in Vietnam. There is no confirmation that 2,3,7,8-TCDD was contained in the dioxin emission reported at the Hempstead facility.

Almost all dioxins are destroyed when exposed to a temperature of 1,472°F for 21 seconds. At higher temperatures, shorter residence times are required. For instance, in an assessment of public health impacts for the New York City Department of Sanitation, it was stated that dioxin levels appear affected by a series of combusion conditions including: (1) minimum of one to two seconds' residence time at 900–1,000°C (1,652–1,832°F), (2) very turbulent conditions in the high-temperature zone, and (3) an air/fuel mixture with sufficient excess of oxygen.[2] These temperatures are roughly the same as those typically found (1,800°F) in a waste-to-energy system.

A statement by EPA in 1981 indicated that dioxin should pose no risk to health given the current rate of emissions.[3] The definitive risk assessment for New York (see footnote 2), which was performed using worst-case assumptions and conservative methods, indicates that cancer risk from dioxin (at the Brooklyn Navy Yard Resource Recovery Facility) "is below levels found by many regulatory agencies to require additional review and probable action to reduce risk."

The environmental fate of dioxin emitted in flue gas has not been established. Although some dioxin emissions may be gaseous, evidence in-

---

2. Fred C. Hart Associates, "Assessment of Potential Public Health Impacts Associated with Predicted Emissions of Polychlorinated Dibenzo-Dioxins and Dibenzo-Furans from the Brooklyn Navy Yard Resource Recovery Facility," August 17, 1984.

3. Interim Evaluation of the Health Risks Associated with Emissions of Tetrachlorodibenzo-*p*-Dioxins from Municipal Waste Resource Recovery Facilities, U.S. EPA, November 1981.

dicates that significant amounts are associated with fly ash or ultrafine particles.[4] This indicates that dioxin may enter the soil through land disposal and the atmosphere by way of gases and particulates which escape control. Dioxins do not migrate substantially in soils.

FEDERAL AIR POLLUTION CONTROL LAWS AND REGULATIONS

Federal air quality requirements that apply to waste-to-energy systems are: National Ambient Air Quality Standards (NAAQS), Prevention of Significant Deterioration (PSD), New Source Performance Standards (NSPS), and National Emission Standards for Hazardous Air Pollutants (NESHAP).

National Ambient Air Quality Standards (NAAQS)

In planning any waste-to-energy project, it is important to consider air standards in the geographical area of the plant. The first step in assessing local air standards is to determine the level of compliance with NAAQS. As directed in Section 109 of the Clean Air Act, the EPA has promulgated national primary and secondary ambient air quality standards for six "criteria" pollutants: sulfur oxides, total suspended particulates, carbon monoxide, photochemical oxidants (ozone), nitrogen dioxide, and lead. These were discussed earlier in this chapter.

Both primary standards "designed to protect public health" and secondary standards (for the protection of public welfare) have been promulgated for the criteria pollutants as shown in Table 8-1. Areas that exceed NAAQS for a pollutant are in nonattainment; those areas that meet or are below the NAAQS criteria for a pollutant are in attainment. An area can be in attainment for some pollutants and nonattainment for others.

The NAAQS levels were set without considering sources of pollutants or availability of control technology. The Clean Air Act, as amended, mandates that each state develop a State Implementation Plan (SIP) for complying with the NAAQS. Ambient air quality is monitored throughout discrete areas called air basins. Data collected from these monitoring programs are used to determine the status of an area (i.e., attainment or nonattainment) with regard to NAAQS.

In a nonattainment area, no new major source of nonattained pollutant can be constructed unless pollution control technology is incorporated that will provide the lowest achievable emission rate (LAER) for that pollutant.

---

4. Fred C. Hart Associates, op. cit.

**Table 8-1. National Ambient Air Quality Standards
for Criteria Pollutants.[a]**

	PRIMARY STANDARD[b] (mg/m^3)	(ppm)	SECONDARY STANDARD[b] (mg/m^3)	(ppm)
Sulfur Oxides (as dioxide)				
Annual arithmetic mean	80	0.03	—	—
24 hour concentration	365[c]	0.14	—	—
3 hour concentration	—	—	1,300[c]	0.5
Suspended Particulates				
Annual geometric mean	75	—	60[d]	—
24 hour concentration	260[c]	—	150[c]	—
Carbon Monoxide				
8 hour concentration	10 mg/m^3	9.0[c]	Same as primary	
1 hour concentration	40 mg/m^3	35.0[c]	Same as primary	
Ozone				
1 hour	235[e]	0.12[e]	Same as primary	
Nitrogen Oxides				
Annual arithmetic mean	100	0.05	Same as primary	
Lead				
Maximum arithmetic mean averaged over a calendar quarter	1.5	—	Same as primary	

[a] Source: 40 CFR 50.4 through 50.12 (July 1, 1980).
[b] Standards are shown in micrograms per cubic meter (mg/m^3) and parts per million (ppm), where meaningful, unless otherwise noted.
[c] Not to be exceeded more than once a year.
[d] A guide for assessing achievement of the 24 hour standard.
[e] Not to be exceeded more than 1 day per year (3 year average).

LAER is defined as the most stringent emission limitation of a pollutant determined by a state or achieved in practice by the class or category of source. To obtain a construction permit in a nonattainment area, pollution offsets, based on the Emission Offset Interpretative Ruling, from other area sources are required in addition to LAER. These measures must actually bring down the net ambient air concentration of the offending pollutant after the new source begins operation. Energy recovery facilities are partially exempted from this offset policy. Under federal regulations, such facilities are required only to obtain the best offsets possible; full offsets are not required, although LAER must still be applied. The issue of whether full offsets are required varies by state.

Prevention of Significant Deterioration (PSD)

Once the attainment/nonattainment status of an area is identified, the applicability of PSD must be determined. PSD regulations only apply to areas that are in attainment or unclassified due to a lack of monitoring data to determine status. New or modified major stationary sources located in a PSD region are required to undergo a preconstruction review and permit process. Twenty-eight major stationary sources that have the potential to emit 100 tons per year (TPY) of regulated pollutants were identified in the 1977 Clean Air Act Amendments. One source is incinerators capable of burning 250 TPD (tons per day) or more of refuse. Waste-to-energy plants capable of firing 250 TPD or more are automatically classified as major stationary sources and required to use best available control technology (BACT) for each pollutant exceeding the 100 TPY level. All other regulated pollutants emitted at rates in excess of specified significant amounts also are subject to BACT, as shown in Table 8-2.

If PSD applies, then a PSD permit is required. There are five major points that a PSD permit application must address, in addition to the use of BACT: (1) ambient air quality analysis; (2) analysis of impacts to soils, vegetation, and visibility; (3) no adverse impacts on a Class 1 area, which includes national parks of more than 5,000 acres and memorials and wilderness areas of

Table 8-2. Significant Emission Levels[a]

POLLUTANT	QUANTITY (TPY)
Carbon Monoxide	100
Nitrogen Oxides	40
Sulfur Dioxide	40
Particulate Matter	25
Ozone	40
Lead	0.6
Asbestos	0.007
Beryllium	0.0004
Mercury	0.1
Vinyl Chloride	1
Fluorides	3
Sulfuric Acid Mist	7
Hydrogen Sulfide (H_2S)	10
Total Reduced Sulfur (including H_2S)	10
Reduced Sulfur Compounds (including H_2S)	10

[a]Source: 40 CFR 51.24(b) (23) (August 7, 1980).

more than 6,000 acres; (4) adequate public participation; and (5) start of construction on time.

New Source Performance Standards (NSPS)

NSPS are designed to control emissions of air pollutants from 28 categories of new or modified sources. Among the regulated source categories are incinerator units capable of burning 50 TPD or more. The applicable standard of 0.08 grain per dry standard cubic foot (gr/dscf) corrected to 12 percent CO_2. This standard applies to individual incinerator units. Therefore, a modular facility with several units that have less than 50 TPD capacity each is required to comply with the less stringent standard of 0.10 gr/dscf.

National Emission Standards for Hazardous Air Pollutants (NESHAP)

The NESHAP were established under Section 112 of the 1970 Clean Air Act. Regulations have been promulgated for four hazardous air pollutants: asbestos, beryllium, mercury, and vinyl chloride. Incinerators have been identified as a source category for beryllium. The applicable standard is 0.022 pound per 24 hour period.[5]

AIR POLLUTION CONTROL EQUIPMENT

The most commonly used pollution control device is a dry electrostatic precipitator (ESP). Other devices include fabric filter systems (baghouses), electrostatic granular filters, and scrubbers.

Electrostatic Precipitators (ESPs)

Electrostatic precipitation involves use of an electrostatic field for precipitating or removing solid or liquid particles from a gas in which particles are entrained. The basic principles involved are particle charging, normally by a discharge electrode, followed by entrapment on a collecting plate of opposite polarity. ESPs are most effective capturing particles in the 1 to 100 micron range.

ESPs represent the most commonly used particulate control device with both European and North American energy recovery facilities. Test reports from European facilities document removal efficiencies of 98 percent and

---

5. 40 CFR 61.32(a) (July 1, 1980).

greater. The lowest outlet grain loading achieved was 0.007 gr/dscf, corrected to 7 percent CO_2. The average loading level was 0.06 gr/dscf (corrected to 12 percent CO_2).

The advantages of ESPs are their relatively high overall efficiencies (associated primarily with removal of larger-sized particles), absence of water requirements, lower pressure drop across the device, and relatively low operating cost.

In the few cases where coal-fired utility boilers with ESPs have been retrofitted to co-fire RDF (Madison Gas & Electric, Wisconsin Electric, and Commonwealth Edison), degradation in precipitator performance was experienced.

Fabric Filters

Fabric filtration uses mechanical screening to trap solid particles. In this type of unit, effluent gas is passed through a porous material (a filter or baghouse installation) that collects particulate matter above a certain size, based on the pore (mesh) size of the filtering material. Particles are initially captured and retained on fibers of the cloth by means of interception, impingement, diffusion, gravitational settling, and electrostatic attraction. The filters are periodically cleaned by mechanical shaking, reverse air flow, or pulse air jets.

Performance tests have shown that smaller-size particles are retained once a filter cake has built up on the filtering surface, in effect decreasing the mesh size. Fibers used as filtering media have included cotton, wool, nylon, Dacron™, and other synthetic materials. The ideal fiber exhibits temperature, acid, and abrasion resistance.

Other factors affecting fabric filter design are size and shape of filters, filter arrangement, spacing of bags, and method of cleaning. Systems are generally efficient for collecting particles greater than 0.5 micrometer; efficiencies over 99 percent are common. If properly designed and operated, consistently high particulate removal efficiencies can be achieved. Tests from the East Bridgewater, Massachusetts facility exhibited 99.8 percent removal efficiency with an outlet loading of 0.001 gr/dscf (corrected to 12 percent CO_2).

Electrostatic Granular Filter

The electrostatic granular filter combines a moving bed filter with an electrostatic grid for high-efficiency collection of particulates. The technology uses the charge occurring naturally on combustion process particulates and subjects them to an electric field which substantially enhances mechanical collection in the packed filter.

The system consists of a cylindrical vessel containing two concentric, cylindrical, louvered screens. The annulus between the screens is filled with a pea-sized gravel medium. Particulate-laden gas enters the filter and travels through the inner screen into the gravel at velocities ranging from 100 to 150 feet per minute. Particulates are removed from the gas stream by contact with the gravel. Clean gas exits through the outer screen and through the appropriate breeching to an exhaust stack. Gas flow may be reversed for some applications.

Electrostatic granular filters have been installed at a number of wood-fired boilers. This control device is used at the 200 TPD mass burn energy recovery facility in Pittsfield, Massachusetts. Data obtained from two test facilities demonstrate a considerable increase in collection efficiency using electrostatic enhancement. Collection of micron-sized particles is especially enhanced with the electric field. These tests have confirmed removal efficiencies in excess of 99 percent.

Scrubbers

Concern about acid gas emissions, which result primarily from burning plastics, has led some states (e.g., California, Connecticut, Massachusetts, and New Jersey) to consider enacting standards for these pollutants. A result has been that projects being developed in these states have proposed to add scrubbers. There are two major types of scrubbers: wet systems and dry systems.

Wet scrubbing systems use a liquid solution to absorb or "scrub" contaminants from flue gas. Scrubber design and characteristics of liquid determine contaminant removal efficiencies. Low- to medium-energy wet scrubbers may reduce fly ash particulates, SO_2, and acid gases, but additional particulate control may be necessary. Wet scrubbers augmented with alkaline reagents may encounter scaling and corrosion, which may be minimized by design changes and use of corrosion resistant materials.

Dry scrubbing involves spray dryers and dry injectors. Atomized droplets of alkaline or caustic slurries react with contaminant gases and form neutral salts such as calcium sulfate ($CaSO_4$), calcium chloride ($CaCl_2$), sodium sulfate (Na_2SO_4), and sodium chloride (NaCl). Dried droplets are collected in particulate control devices. Control efficiencies for SO_2 and HCl are in the ranges 70–90 percent and 80–98 percent, respectively. Removal efficiency for SO_2 may be dramatically increased by addition of fabric filters at the back end of spray dryer systems.

SOURCES AND CONTROL OF WASTEWATER DISCHARGE

Sources of wastewater from mass burning and RDF dedicated boiler combustion systems are:

- Continuous and intermittent blowdown
- Equipment and facility washdown
- Pretreatment filter backwater
- Demineralizer-neutralized regenerate
- Quench water
- Site drainage
- Sanitary water

The first two wastewater sources are used to provide water needed for the ash quench tank. The quench system is used to cool incinerated bottom ash/residue and fly ash, which are typically conveyed to the quench tank. Ash and residue in the quench tank are removed by a drag-line conveyor.

Depending on the approach used at a specific facility, clean wastewater (blowdown, pretreatment filter backwater, demineralizer-neutralized regenerate) may be used as washdown water. In any case, washdown water is piped to a sump in order to settle large solids before pumping to the quench tank.

Water loss in the quench tank is a result of two conditions: (1) evaporation and (2) absorption by ash/residue. Hot bottom ash/residue (700 to 800°F) causes water in the quench tank to remain at a relatively high temperature, resulting in some loss due to evaporation. Actual quantity lost will depend on the amount of debris floating on the surface. A high percentage of floating debris reduces the surface area of the quench tank, thereby limiting evaporation.

In a mass burning facility, considerable ash quench water is lost through absorption by the cooling ash and residue by-products (about 30 percent by weight of the ash/residue discharged). For example, water losses due to absorption might range from about 2,500 gallons per day (GPD) for a 100 TPD mass burn plant up to 12,500 GPD for a 500 TPD plant. Water content of ash from an RDF dedicated boiler is about 50 percent on a weight basis. Typical quantities range from about 1,800 GPD for a 100 TPD RDF dedicated boiler to 9,000 GPD for a 500 TPD boiler.

Wastewater from boiler blowdown is insufficient to provide all makeup water needed by the quench tank. The remainder may be supplied by washdown water, if obtained from a source other than blowdown. Two other possible sources of washdown and supplemental water to the quench tank are: (1) pretreatment filter backwash and (2) demineralizer-neutralized regenerate. These wastewaters either can be used for in-plant purposes or can be discharged to a sanitary sewer.

To control buildup of suspended and dissolved solids, water in the quench tank typically is circulated to a clarifier. Part of this circulation system may include a tank for storage of excess water that might enter the quench system. This eliminates the need to discharge water from the quench tank.

Drainage of free water from ash stored at the plant site is another source of wastewater in mass burning systems without residue recovery and RDF dedicated boiler units. Ash/residue is discharged from the quench tank into a bin or truck. Free water drains from the ash/residue while stored. Drainage of free water can be improved by tilting the storage container. Maximum drainage of free water at the plant site will reduce the amount that will leak during transit of ash/residue to a disposal site. A drain located near the ash/residue discharge point supplies water to the circulation system.

Site drainage water should be kept separate from other wastewater and waste receiving, ash handling, and other activities, so that it can be sent directly to a wastewater system. Sanitary water used in-plant also is piped directly to a public sewerage system.

The amount of wastewater produced by a boiler under co-fire conditions should be similar to the quantity discarded prior to the use of RDF. Co-firing of RDF will cause a change in parameters (physical, chemical, and biological) of bottom ash transport water. Under proper conditions, however, these changes should be insignificant.

Sources of wastewater at an RDF production facility are:

- Equipment and facility washdown
- Site drainage
- Sanitary water

Equipment and facility are cleaned at least once per week. If the production facility is located at a site separate from the combustion unit, the plant probably will be swept. A production facility sited adjacent to the combustion unit will have an available source of wastewater, such as blowdown, for washing. Without such a supply, a production facility must purchase water and possibly provide for settling of solids in the washdown water. Washdown, site drainage, and sanitary water may be discharged directly to a sanitary or storm sewer, where applicable.

FEDERAL LAWS AND REGULATIONS

Through the Federal Water Pollution Control Act of 1956, the Water Quality Act of 1965, the Clean Water Act of 1977, and related amendments, the federal government has attempted to restore and maintain water quality. Energy recovery facilities are required to comply with federal standards regulating wastewater discharge. Typically, discharge will be to a public wastewater treatment plant. In such cases, the National Pretreatment Standards apply. However, if discharge is to a natural water body, then a National Pollutant Discharge Elimination System permit is required. As a minimum, the National Pretreatment Standards must be met. Depending on quality of the body of water, more stringent requirements may be imposed.

National Pollutant Discharge Elimination System (NPDES)

The EPA administers a process by which each source discharging into a navigable water must apply for an NPDES permit containing specific limitations on various constituents in its discharge. The NPDES permit sets effluent limitations on the basis of national effluent guidelines and ambient standards. Through Section 306 of the Clean Water Act, EPA sets effluent limitations for new sources. As a result, national standards of performance for at least 27 specified industrial categories have been established. One of these categories is steam electric power generators, including solid waste energy recovery facilities that generate electricity. Section 306 defines the term "standard of performance" as a standard for control of pollutant discharge which reflects the greatest degree of effluent reduction achievable through application of best available demonstrated control technology (BADCT), process operating method, or a suitable alternative.

For the steam electric power generating industry, EPA has identified categories of typical industrial operations and (for each such category) types of technologies available to reduce the discharge of pollutants. Discharge levels associated with such technologies became the basis for "effluent limitations" for this industry; see Table 8-3. These effluent limitations are somewhat flexible in that they are subject to change based on information supplied by the applicant specific to its technology or process. After effluent guidelines are imposed, EPA determines whether relevant ambient water quality standards will be met before issuing an NPDES permit.

The NPDES permit process begins when a particular discharger applies for a permit. At this time, effluent guidelines are imposed. Then an assessment is made to determine if relevant ambient water quality standards and policy goals would be attained if the effluent limitations shown in Table 8–3 were applied. If these objectives cannot be attained, then effluent limitations are made stringent enough to attain ambient standards. Once it has been determined that the ambient standards can be met, an NPDES permit is issued.

National Pretreatment Standards

If an energy recovery facility discharges into a publicly owned treatment works (POTW), the major water quality standards that must be met are the National Pretreatment Standards. These standards are based on requirements of the Federal Water Pollution Control Act as amended by the Clean Water Act of 1977 (P.L. 95-217). The Clean Water Act established responsibilities of federal, state, and local governments, industry, and the public to implement standards that would control pollutants which pass through, or interfere with, the treatment processes of publicly owned wastewater treatment facilities or which may contaminate sewage sludge.

The design of each wastewater treatment facility must limit pollutants according to the facility's particular situation and incorporate these limitations

Table 8–3. Pretreatment Guidelines for Steam Electric Power Generation Facilities Discharging into Sewerage Systems.[a]

EFFLUENT SOURCE	EFFLUENT CHARACTERISTIC	DAILY MAXIMUM STANDARD (mg/l)[b]	AVERAGE MONTHLY STANDARD (mg/l)[c]
Low Volume Waste Sources	Total suspended solids	100	30
	Oil and grease	20	15
Bottom Ash Transport Water[d]	Total suspended solids	100	30
	Oil and grease	20	15
Metal Cleaning Wastes	Total suspended solids	100	30
	Oil and grease	20	15
	Copper, total	1	1
	Iron, total	1	1
Boiler Blowdown	Total suspended solids	100	30
	Oil and grease	20	15
	Copper, total	1	1
	Iron, total	1	1
Cooling Tower Blowdown	Corrosion inhibitors[e]	No detectable amount	No detectable amount

Table 8–3. (continued)

EFFLUENT SOURCE	EFFLUENT CHARACTERISTIC	MAXIMUM CONCENTRATION (mg/l)[b]	AVERAGE CONCENTRATION (mg/l)[b]
Once-through Cooling Water	Free available chlorine	0.5	0.2
Cooling Tower Blowdown	Free available chlorine	0.5	0.2

		OTHER STANDARDS	
All discharges, except Once-through Cooling Water	pH	6.0–9.0	
All Discharges	Polychlorinated Biphenyl compounds (PCB)	None	
Fly Ash Transport Water	Total suspended solids Oil and grease	None None	
Any Units	Free available chlorine Total residual chlorine	No unit may discharge for more than 2 hours in any one day, and not more than one unit in any plant may discharge at any one time	
Blowdown from Recirculated Cooling Water and Blowdown from Cooling Ponds	Heat	Temperature must not exceed the lowest temperature of recirculated cooling water prior to addition of makeup water	
Main Condensers	Heat	None, except for above	

[a] Source: Based on 40 CFR 423.10–426.26 Steam Electric Power Generating Point Source Category, Small Unit (less than 25 MW unit or less than 150 MW system) subcategory.
[b] Multiply this concentration times amount of wastes from a particular effluent source to establish standard of performance (mg/l = milligrams per liter).
[c] Average of daily values for 30 consecutive days shall not exceed these limits.
[d] For these concentration standards, divide product of multiplication by 20 to establish standard.
[e] Materials added for corrosion inhibition include zinc, chromium, phosphorus, and others.

into the applicant's NPDES permit. As an example, in some cases, heat may be beneficial and accelerate the effectiveness of a wastewater treatment process, particularly in cold weather or cold climates. However, because the average wastewater treatment facility includes biological processes that could be damaged above 104°F, a national standard was established to set a baseline for local standards to be developed which limits thermal discharge to a wastewater treatment facility.

"General Pretreatment Regulations for Existing and New Sources of Pollution" are codified in 40 CFR 403, with specific regulations affecting "Steam Electric Power Generating Point Sources" established in 40 CFR 423. These standards include general discharge prohibitions which apply to all nondomestic discharges into wastewater treatment facilities. Standards for new sources such as waste-to-energy facilities are based upon the best available demonstrated technology economically achievable.

For energy recovery facilities that generate electricity, the "Small Unit Subcategory of the Steam Electric Power Generating Point Source Category" of the National Pretreatment Standards applies. Pretreatment standards delineated in 40 CFR 423.21 are for "any small industrial facility primarily engaged in generation of electricity (less than 25 megawatts per unit of less than 150 megawatts per system) or distribution and sale which results primarily from a process using fossil-type fuel (coal, oil or gas), or nuclear fuel in conjunction with a thermal cycle employing the steamwater system as the thermodynamic medium." Since waste-fired power facilities fall into this size range and use a steam-water system, these standards apply. These specific industry pretreatment standards are generally the same effluent limitations as those listed in Table 8–3.

The NPDES permit program and/or the National Pretreatment Standards programs may be administered by either federal or state agencies. The appropriate state agency may develop and submit a plan to the U.S. EPA and, pending approval, assume the authority for these programs.

Residue and Ash

On average, 24.4 percent of municipal solid waste is ash and residue. A breakdown yields about 18.3 percent noncombustibles (e.g., glass, metals) and 6.1 percent ash.

In a mass burning system, the quantity of ash and residue generated is about 24.3 percent of the waste burned. The 0.1 percent difference from the quantity listed above is the amount of particulates lost to the atmosphere. The type of mass burning system determines the character of the ash and residue generated. Direct combustion systems (waterwall and refractory-lined incinerators) generate more fly ash, while modular systems produce more bottom ash/residue. In modular incinerators which operate as gasifiers, there is insufficient air for complete combustion of volatile gases in the primary chamber. This means less turbulence from the air and, therefore, less entrainment of ash particles in combustion gases. Direct combustion systems are designed for complete combustion of volatile gases in the incinerator. Excess air needed to insure complete combustion results in ash particles carried along with hot gases.

For comparative purposes, a breakdown of the residue between direct combustion and modular units is given here:

| | PERCENT BY WEIGHT | | |
	BOTTOM ASH/ RESIDUE	FLY ASH	TOTAL
Direct Combustion	20.2	4.1	24.3
Modular Incineration	24.3	0.3	24.3

This points out the need for reliable air pollution control equipment for direct combustion systems.

The noncombustible portion of residue from RDF systems typically amounts to 25 percent of waste processed. This fraction of the waste stream includes materials such as metal and glass as well as large combustibles such as wood. Separation of waste into two parts, one containing only combustibles and the other noncombustibles, is very difficult, and there is an overlap between the two fractions of processed waste. RDF has an average ash/residue content of 7.3 percent. As with mass burning systems, there is a difference between RDF combustion processes regarding ash production. About 16 percent fly ash is generated in a dedicated boiler (spreader stoker), with the remainder bottom ash/residue. RDF burned in a suspension-fired unit, typically co-fired, has 50 percent fly ash and the remainder bottom ash/residue. Again, the form in which ash is produced has an effect on the air pollution control system. This is particularly important in a co-fire application using an existing facility. Ash from RDF may have a negative impact on operation of air pollution control equipment in place.

Ash/residue found to be nonhazardous can be disposed at a landfill that accepts other types of nonhazardous waste. In some states, the residue can be put on the landfill without daily cover as long as the putrescible content (e.g., food waste) has been adequately destroyed. Cover material is used to control rodents and flies as well as to prevent blowing paper and fires. The fine particle size of ash makes erosion from runoff a potential problem if no cover material is used. Periodic covering can control this potential problem.

Resource Conservation and Recovery Act

In an effort to protect our land from unsafe solid waste and hazardous waste disposal practices, the Resource Conservation and Recovery Act (RCRA) was passed in 1976. The major programs of RCRA created a new federal hazardous waste regulatory program and prohibited open dumping of solid wastes. Hazardous waste regulations, authorized by Subtitle C of RCRA, were promulgated by EPA on May 19, 1980. Municipal solid waste-to-energy facilities are covered by these regulations, but most facilities are not significantly affected. The only facilities largely affected are those which accept hazardous waste from small-quantity generators and/or those having hazardous process residue. These facilities must dispose residue in accordance with RCRA regulations.

Resource recovery facilities that accept only residential waste are excluded from the regulations (40 CFR 260) because these wastes are specifically identified as nonhazardous. However, wastes from small businesses, offices, and industries are subject to regulation and, therefore, must be tested to determine if they are classified as hazardous under the regulations. A waste is considered hazardous if it is ignitable, corrosive, reactive, or toxic, or if it is listed in the regulations. Residues from resource recovery facilities are not listed as hazardous wastes. (In some states, however, residues are classified as "special" or "hazardous" and are required to be disposed in facilities with more extensive pollution control provisions than those receiving municipal solid waste only.)

Toxicity is the characteristic of greatest concern with waste-to-energy facility residues, including noncombustible materials derived from RDF production. The extraction procedure (EP) is designed to identify those toxics which might be leached from a landfill in quantities that could pollute a ground or surface water source. EP toxicity tests were performed on residues from six waste combustion systems of various designs. Residues sampled were fly ash, bottom ash (quenched or dry), combined ash, and air classifier heavy fraction. The only residue which failed the EP toxicity test was the fly ash from three of the plants. When the fly ash and bottom ash streams were mixed, which is performed automatically within some process units, all of the combined ash samples passed the test.

Thus, when ash streams are mixed in the process unit, compliance with regulations may be achieved. When fly ash is treated as a separate waste stream, it may require special treatment and/or disposal. However, fly ash is considered nonhazardous until tested and proven to be hazardous. Facilities may apply for a hazardous waste treatment permit to mix fly ash with bottom ash outside the unit, petition for an exclusion, modify the facility to effect mixing within the process unit, or dispose of ash at an approved hazardous waste disposal site.

It is the responsibility of the owner/operator of an energy recovery facility to determine whether residues are hazardous under the new regulations. If the residues are hazardous, the owner/operator must notify EPA and comply with either the requirements for a hazardous waste generator or the requirements for an owner or operator of a hazardous waste treatment, storage, or disposal facility. However, these hazardous waste regulations probably will not have a significant impact on energy recovery facilities.

Under Subtitle D of RCRA, there is a discussion of handling solid waste which is not hazardous. Of greatest impact to energy recovery facilities is the portion of Subtitle D concerned with open dumps. EPA has developed criteria to distinguish between open dumps and sanitary landfills. Solid wastes not subject to hazardous waste regulations must be used for resource recovery, disposed in sanitary landfills, or otherwise disposed in an environmentally sound manner. These requirements must be complied with when disposing of nonhazardous residue from any resource recovery facility.

STATE ENVIRONMENTAL STANDARDS

Environmental standards for most states are designed after federal regulations. However, at times they may be tempered by local or site-specific considerations. It would not be practical to review the regulations for all states (or other local entities) in this chapter. Therefore, some air and water pollution controls as well as solid waste regulations are reviewed for selected states in order to provide examples.

Certain regulations may differ from state to state. For example, the California Department of Health Services considers fly ash hazardous until declassified by administrative procedure or a waste extraction test. Once declassified, other hazardous waste regulations do not apply. New York State regulations, on the other hand, do not define ash as hazardous. Texas, Pennsylvania, and Florida also have determined ash to be nonhazardous.

Guidelines on emission limits for resource recovery facilities in California are shown in Table 8-4. In 1983, California adopted a PM_{10} ambient air quality standard designed to limit concentration of particles less than 10 microns in diameter:[6]

24 hour standard = 50 micrograms per cubic meter
Annual geometric mean = 30 micrograms per cubic meter

California also has established a procedure whereby toxic air contaminants (non–criteria pollutants) are identified. The state will subsequently develop and adopt control measures.

---

6. The EPA also is considering a PM_{10} standard. See *Federal Register* (Part II, 40 CFR, Parts 50, 53, and 58) (March 20, 1984).

**Table 8-4. Guidelines on Emission Limits for
Resource Recovery Facilities.[a]**

POLLUTANT (8 hour average)	EMISSION LIMITS (ppm at 12% CO_2)
SO_2	30
NO_x (as NO_2)	140–200
CO	400
THC (as CH_4)	70
HCl	30
TSP[b] (total)	0.01 gr/dscf at 12% CO_2
(less than 2 microns)	0.008 gr/dscf at 12% CO_2

Additionally, to ensure control of organics, a temperature of
980 ± 110°C (1,800 ± 200°F) and a residence time no less than
1 second should be maintained in combustor.

[a]Source: "Air Pollution Control at Resource Recovery Facilities," State
of California Air Resources Board, May 24, 1984.
[b]Based on average of three consecutive EPA Method 5 tests at full boiler
capacity. For the purpose of this guideline, emission limits for particulate
matter are based on solid particles contained in the EPA Method 5
measurements (front half). However, some districts include the front and
back portions in determining the particulate matter emissions. For offset
purposes, both the solid and the condensable fractions are considered as
particulate matter in most California air pollution control districts.

The particulate emission limitation established in Connecticut is essentially the same as California (0.01 gr/dscf). The states of Maryland and New Jersey have adopted an emission standard of 0.03 gr/dscf.

In Connecticut, the maximum emission of Pb must not exceed NAAQS (1.5 micrograms per cubic meter). With regard to other heavy metals, ambient effects must be evaluated for nickel (Ni), chromium (Cr), cadmium (Cd), arsenic (As), Hg, and Be.

Finally, even within states, regulations can differ. For example, primary responsibility for controlling stationary sources in California rests with local air pollution control or air quality management districts, of which there are 40. Thus, such districts are authorized to establish stricter emission standards than those set by law.[7]

SUMMARY

Solid waste energy recovery can be viewed as an environmental benefit in that the amount of material that requires disposal is reduced. In addition,

7. "Air Pollution Control at Resource Recovery Facilities," State of California Air Resources Board, May 24, 1984.

there is the potential added benefit of energy production. The situation does exist, however, in which environmental benefits can be offset by emissions from a facility.

Combustion of solid waste or RDF produces a variety of pollutants that require control. Federal and state laws and regulations exist that restrict discharge of flue gas, wastewater, and ash/residue.

Characteristics of flue gas depend on various factors including composition of feed and process design. Most air pollutants are produced at very small rates. However, due to their nature, they exhibit the greatest potential for adverse impact on the environment. National standards have been established for most pollutants identified in flue gas.

Control of air pollutants is provided by such devices as the electrostatic precipitator, fabric filter system (baghouse), electrostatic granular filter, and scrubber. These devices are sophisticated and exhibit high removal efficiencies.

Process wastewaters, on the other hand, tend to be recirculated in-plant to the extent that in many cases, the need to purchase makeup water is reduced. In cases where wastewater is discharged, settling, clarification, and/or treatment before discharge into a natural waterway or wastewater treatment plant can be easily and effectively achieved.

Tests performed on process residues and ash from operating municipal waste-to-energy facilities have shown these materials to be generally nonhazardous. Thus, the majority of operating facilities dispose of ash and residue by-products at sanitary landfills without processing and/or special handling. Special disposal requirements for ash and residue have not been the practice to date; however, such requirements may need to be met if problems are identified on the basis of tests conducted on a project-by-project basis.

Federal and state laws and regulations exist that restrict wastewater discharge and ash/residue disposal produced as by-products from municipal waste-to-energy systems. However, in practice these regulations have not resulted in elaborate, costly control strategies. Project developers should recognize, however, the differences in restrictions from state to state and should take these into consideration in project planning.

9
PROJECT DEVELOPMENT

The focus of this chapter is on the risks inherent in small-scale energy recovery projects and the manner in which these risks transcend alternative ownership, operations, procurement, and financing approaches that can be applied in developing these systems. The development and procurement process is affected by advisors involved in the projects, especially if an advisor has a design engineering interest.

PAST EXAMPLES

A review of how small-scale energy recovery projects have been implemented in the past provides a basis for future projects. Table 9-1 is a compilation of selected U.S. small-scale projects. Early projects were typically procured using the traditional architect/engineer (A/E) method that is most often used for public works oriented capital acquisitions. Some communities, such as Blytheville, Arkansas and Salem, Virginia, used general obligation debt to pay for their facility. In Norfolk, Virginia, military construction funds were applied, as it is a U.S. Navy facility. The City of North Little Rock, Arkansas used municipal revenue bonds to finance its plant. Communities developing plants later diverged from this traditional procurement approach: in Auburn, Maine, a modified turnkey contractor was hired; in Hampton, Virginia, a turnkey contractor was awarded the job; and in Pittsfield, Massachusetts, a full-service, privately owned/operated project was put in place. Still the A/E public debt approach often prevails, but procurement through private operators becomes more prevalent and difficult. Projects in Glen Cove, New York; Oswego County, New York; Oneida County, New York; Miami, Oklahoma; Tuscaloosa, Alabama; Pascagoula, Mississippi; Portsmouth, New Hampshire; Cattaraugus County, New York, and elsewhere reveal this trend.

A variety of procurement and financing approaches has resulted in differing allocations of risk. Oftentimes, these allocations are a by-product of a predetermined procurement and financing choice. Risk allocation should be performed as input to these decisions and, if done objectively, will result in a better-defined project and in all parties being prepared for (versus surprised by) their long-term responsibilities, both financial and administrative.

STAGES OF A PROJECT

The stages of developing a project are: feasibility, procurement planning, procurement, design-construction, and operations. These stages have spanned three to ten years from planning to operations. The usual should be four to five years from the time the concept is brought forward as a waste disposal alternative.

To get started with a project, two important elements must be in place. First, a garbage disposal crisis has to be present or imminent. A crisis can manifest itself in the following manner: there is diminishing landfill capacity; new landfill regulations will cost too much to put into place; costs for a new landfill will only escalate higher and now is the time to do something; or a new landfill siting has been voted down and politicians are wary of trying again. As despondent as local government may be about the media, the media can be a project's biggest ally in communicating a garbage crisis. The media must be utilized fully and carefully to make local constituents aware of the choices. Nothing can focus attention on energy recovery like strong landfill regulatory actions. In the 1970s, energy recovery promoters declared, "Regulate those landfills properly, and energy recovery will take care of itself." By the 1980s, states have developed stronger enforcement capabilities but often still lack the "hammer" to close a polluting landfill. Regardless, their continued regulatory efforts will help focus on the need for an option. There is also nothing like a landfill siting evaluation to get nearby residents stirred up. This is not to imply that siting an energy recovery facility is easy, but landfills can be even tougher.

Second, political leadership has to be present to move forward. Without it, the best laid plans collect dust on a planner's bookshelf. Political leadership is either present or not. There is little that can be done to get it if it is not there. If it is not present, wait until it is or until the waste disposal crisis comes to the point where the leadership has the fortitude to step forward. History shows that a good idea is only good if the timing is right.

The planning and development process most often is conducted with the assistance of a consultant or group of advisors. With the selection of advisor(s) comes a predetermined set of biases, based upon experience as well as interests. Project managers will do well to understand any technology, procurement, or financing preferences advisors may have. Their particular biases will affect the advice rendered and, as a result, the direction a project may take.

Typically, these efforts start with a feasibility study. Project development agencies have spent from zero to several hundred thousand dollars on such studies. The study's objective should be to focus on an energy recovery approach that can be implemented and will be competitive with anticipated

Table 9-1. Review of Implemented Small-Scale Energy Recovery Projects.

PLANT	SIZE (TPD)	PROCUREMENT APPROACH	FINANCING METHOD	CONSTRUCTION	OPERATION	ALLOCATION OF MAJOR RISKS		
						MARKETS[a]	WASTE STREAM[b]	CHANGE IN LAWS AND REGULATIONS
Lamprey Cooperative, University of New Hampshire, Durham, New Hampshire	108	Turnkey	$3.5 M (1979) • G.O. bonds	Cooperative	Cooperative	Cooperative/University agreement	Cooperative	Cooperative
N. Little Rock, Arkansas	100	A/E	$2.3 M (1980) • G.O. bonds • Municipal revenue bonds • City general fund	Consumat	City then Consumat (facility closed November 1983)	City/Koppers 20-year contract	Municipal control and economics	City
Windham, Connecticut	108	Turnkey	$3.7 M (1980) • DEP grant • FHA loan	Consumat	Town	Town	Agreements between Town and participating communities	City
St. John's University Collegeville, Minnesota	65	Turnkey	$2.3 M (1977) • U.S. HUD loan (60%) • U.S. DOE grant (2%) • In-house debt (38%)	Basic	University	University	Contracts involving University, a municipality, and private haulers	University
Auburn, Maine	200	Modified turnkey	$4.7 M (1981) • G.O. bonds (74%) • State grants (10%)	Consumat/Global via construction agreement with City	Consumat (3 years)	City	City through put-or-pay contracts	City

Location	Capacity	Approach	Financing					
Hampton, Virginia	200	Turnkey	• U.S. DOE grant (6%) • Interest on (10%) G.O. bonds $10.4 M (1979) • municipal revenue bonds (67%) • NASA (33%)	NASA	City	20 year contract between City and NASA	Economics	City
Portsmouth, New Hampshire	200	Modified full-service approach	$5.8 M (1980) • G.O. bonds	Consumat	Consumat	City/Pease Air Force Base 10-year contract	City/towns	City
Gallatin/Sumner County, Tennessee	200	A/E	$9.8 M • Grants, TVA low-interest loans (38%) • 3 year construction loan (62%) • Municipal revenue bonds (1983)	Contractor	Authority	Contracts between Authority and 4 energy markets	Authority contracts with Sumner County waste suppliers	Authority
Salem, Virginia	100	A/E	$1.9 M (1977) • G.O. bonds	Consumat	City	City/Mohawk Rubber contract	City	City

[a] Energy market contracts typically 15 to 20 years, "take-or-pay."
[b] Waste stream contracts typically 15 to 20 years, "put-or-pay."

195

landfill costs. The study can provide the focus for a better approach, but by no means should it be viewed as the final word on almost any aspect of the project that may evolve. The real definition of a project and the determination of where resources are best applied result from the project's procurement, development, and contracting.

PROJECT BUILDING BLOCKS

There are several factors that are essential in the development of a resource recovery project. These can be viewed as key elements or "building blocks." If any one of them is missing, it is unlikely that a project can be procured, much less implemented (see Figure 9–1). These building blocks include:

- Energy market(s)
- Waste supply
- Facility site
- Landfill for residue or use in emergency
- Project sponsor
- Front-end resources (staff and funds)
- Contractor(s) to implement the project
- Sources of capital financing

The most critical elements are the energy market and the waste supply. They relate to the project's economics which, in itself, will usually be the most important factor in deciding whether to proceed with implementation. The other factors, however, are more qualitative in nature and are sometimes overlooked by communities who believe that a project can be developed if the economic feasibility is evident. A brief review of these other factors will give an appreciation of their importance and the need to consider them in the overall feasibility assessment process.

Securing an energy market should be top priority in development efforts. Be it for waste-derived steam, electricity, or fuels, purchasing alternative energy is usually a new activity for the market. Much hand-holding and education will need to take place at the operating, as well as different management, levels of the energy market. Visiting operating plants provides assurance that the concept works. Negotiating letters of intent which pin down responsibilities of various parties and clarify the contract posture related to term, pricing formula, providing land for facility site, interest in ownership and operations, etc., is important to the market commitment process. If properly done, the energy market's interest will be committed and will provide well-defined input to the project's technical concept and facility performance requirements.

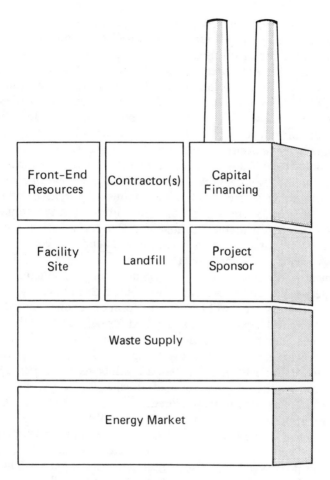

Figure 9-1. Project building blocks.

Without a raw material to process, an energy recovery facility has no means of surviving. Waste supply has to be logistically, contractually, and legally made available to a project. There may be a need to establish a transfer station system for efficient deliveries. Communities will need to know what the extra costs will be to deliver waste to the new project. Usually, long-term waste supply contracts must be established for project financing, unless the waste is already collected and controlled by the project sponsor. This can be a difficult process and one that may require special counsel and even state authorizing legislation. Oftentimes, waste districts or authorities are formed that have special powers which ease the process of waste supply control, contracting, and rate covenants for the service. The local project manager must establish preliminary commitments such as resolutions and contract forms that are akin to the long-term waste supply agreements ultimately executed.

Obtaining a site on which to build the facility is a critical project component. The difficulties in securing a site and the time required to do so often are not fully appreciated. This aspect of the project must be addressed early, and resolved unequivocally, well before implementation begins. Extensive siting studies may be desired to pick the "best" site. Experience has shown that the "best" technical site may be far from the "best" politically acceptable site. A politically acceptable site should be sought without commissioning extensive siting studies. Obtaining options to purchase the site and developing a strategy for site acquisition and permitting should be accomplished before procurement commences. Permitting activities should be initiated at the same time as procurement and be completed once technology/vendors have been selected.

Implementing a waste-to-energy project does not eliminate the need for a landfill. Between 30 and 50 percent by weight and 5 to 15 percent by volume of the input to the system will leave in the form of residue that will require landfilling. Landfill capacity also is required for periods when the system is down for scheduled or unscheduled maintenance, as well as for waste that cannot be processed for energy recovery, e.g., certain bulky wastes and construction debris. An existing or new landfill will be required. Perhaps a participating community can provide landfill capacity, or a separate private landfill service may need to be procured.

A project sponsor must step forward. This refers to the need for an individual or agency to accept responsibility for developing and implementing the project. There must be a single, committed coordinator to keep the development process moving. The project manager assigned will need to be an active spokesperson with management abilities as well as the ability to communicate effectively and gain widespread support. Depending on the ownership and financing approach of a particular project, the coordinating role may be played by a municipality, special authority, industry, utility, or system vendor. In all cases, without such a sponsor, inevitable technical pitfalls and institutional barriers will unnecessarily delay or terminate the project.

Front-end resources for planning and development efforts, before the contractual commitments have been obtained and financing has been secured, are often difficult of obtain. Nevertheless, front-end money for consultants and time for inhouse staff are both critical to a project's ultimate success. In planning and developing an energy recovery project, the funds needed up front (before the project is financed and construction commences) could amount to several hundred thousand dollars.

Once the preceding project building blocks have been defined, contractor(s) will need to be selected for purposes of financing, design, construction, and operations. As discussed later, contractors for these purposes can be procured separately or as a full service, depending upon the project administrative structure desired. Regardless, contractors should be selected based upon their experience and track record in providing services/equipment for similar projects, their competitiveness, overall approach to the problem, and ability to provide financial resources and/or guarantees necessary to support the project's financing. Too often, projects have been implemented without giving early and full consideration to the abilities of the contractor(s) selected. Selecting inappropriate contractors, particularly those with less proven technology and insufficient financial strength, will lead to stalled or terminated implementation or to the project sponsor's assuming greater levels of risk and financial exposure than originally contemplated or desired.

The last building block is a source of capital for the project provided through a sound financing method. In most future projects, this will involve some type of revenue bond financing, perhaps combined with private equity. It can take considerable time to get projects to the point at which bonds can be issued. Moreover, the ability to finance a project is dependent on the other building blocks. They form the security structure for the project. The use of "project financing" to finance small-scale energy recovery projects has been especially difficult and is discussed in Chapter 7.

INSTITUTIONAL SETTING

The institutional environment of a project is sometimes difficult to gauge. In any project there are a number of parties-at-interest who will have an impact on decision making. If there is a solid waste crisis (i.e., little or no landfill capacity remaining, or lack of disposal outlets with which a contractual relationship can be arranged), then the sentiment to plan and develop an energy recovery project quickly may be widespread among municipal officials, solid waste collectors, the general public, interest groups, and other local parties. On the other hand, if local landfill capacity is not critically limited, disposal costs are not exorbitant, and the site location is not overly distant, then local interest and momentum for resource recovery may not be strong, even if an energy recovery project appears economically attractive. Oftentimes, energy recovery projects that were otherwise economically feasible have been abandoned in the planning stages simply because sufficient support could not be generated among the local leaders.

The planner must carefully review the sentiment in the community for energy recovery and must assess the posture and philosophy of the municipal officials, the local haulers, local markets, the general public, and others. Local haulers who would have to haul to a more distant location could not be expected initially, if at all, to offer strong support. A private landfill operator whose operation could be affected adversely would also not be expected to provide support. Other institutional barriers in the form of competing demands for planning monies or staff time on other local projects may sidetrack efforts away from energy recovery. The existence of a potential implementation agency such as a solid waste authority or perhaps ongoing efforts toward some other solid waste project could affect the political climate and interest for a resource recovery project. These and other factors bear upon the institutional environment for the project. Essentially, it must be a "politically feasible" undertaking as well as economically justified. The planner will have to sound this out over the course of the feasibility assessment.

A small-scale solid waste energy recovery facility is a capital intensive project. For example, bonded debt on a 200 TPD (tons per day) plant could exceed $15 million dollars. Most communities in the United States have limited capital to apply to such projects, and their borrowing capacity may be at or near its legal limit.

Today there is a trend toward revenue bond financing and full-service contracting with a private system vendor for resource recovery projects, in view of the substantial costs and risks involved. Fewer projects are being financed through general obligation debt secured by the backing of the taxpayers. Typically, projects are financed through a special authority such as a solid waste authority or industrial development authority empowered to issue revenue bonds secured by the project revenues and assets. Such financing may or may not include private equity. In these forms of financing, and with private participation through the vehicle of a special financing entity, many of the capital limitations and risks are removed from the municipality, although it would be obligated to supply solid waste to the project.

In analyzing the potential for developing a successful project, it is appropriate to determine what positive and negative conditions exist which critically affect ultimate feasibility. This review can be completed through the use of a checklist. Table 9–2 presents a checklist that can help evaluate the project potential. If substantially more negative than positive entries are made on the checklist, it is doubtful that an energy recovery project is feasible, at least not in the near term.

Table 9–2. Evaluation Checklist.

ENTITY	POTENTIAL ROLES
The Municipality	Owner and/or operator of transfer station(s), waste supplier for publicly controlled waste to the project, source for general obligation debt, project sponsor, operator of the energy recovery facility, procurer and/or provider of landfill for residue disposal and backup, procurer and provider of transfer station(s).
Electric Utility	Purchaser of cogenerated electricity, owner and/or operator of energy recovery facility, owner and/or operator of transfer station(s).
Other Incorporated Municipalities	Waste suppliers to the project. Providers of landfill and/or transfer stations or sites for them.
Federal and State Regulatory Agencies	Provider of permits and perhaps technical and/or financing assistance.
Steam or Electricity Customer	Steam or electricity purchaser, provider of energy recovery facility site, operator of energy recovery facility, owner of energy recovery facility.
Private Waste Haulers	Waste supply contract or, transfer station(s) operator.
Private Landfill Owner and Operator	Provider of landfill for residue disposal and backup.
Contractor(s)	Designer, builder, operator; owner of energy recovery facility and/or transfer station(s).
Outside Party(ies)	Owner, bondholders.
Federal and State Governments	Provider of capital for facilities, if available; technical assistance; siting assistance.

ROLES THAT CAN BE TAKEN

The implementation of an energy recovery project will need to include the following physical elements:

- A *waste delivery system,* including transfer stations, to get refuse to the energy recovery facility
- An *energy recovery facility*[1] to process the refuse and generate steam for delivery and use by the steam customer and/or electricity for sale to the utility and residue for haul (or preprocessing prior to haul) to a disposal site

---

1. For processed fuel systems, it is possible to separate the fuel processing from the combustion process.

- A *sanitary landfill* to accept the residue and, if necessary, serve as an emergency disposal site for refuse in the event that the energy recovery facility is inoperable for an extended period (greater than two to three days)

The different entities that could be involved in a project, and the role(s) each party could play are listed in Table 9–2.

Before a project can be implemented, the sponsor will need to determine which respective roles it would prefer taking in such a project. In so doing, the objectives of each party have to be analyzed with respect to the risks that are assumed in taking the respective role(s). Common objectives for this type of project include:

- Having control over the project's development and implementation
- Having control over the facility's operations and associated costs
- Having a fair degree of confidence in the cost for participating in the project
- Passing off major design, construction, and operation risks to other parties better equipped to deal with them
- Being able to take advantage of significant funds that may be available from the federal or state governments
- Being assured of a long-term, reliable, environmentally sound, and economical source of solid waste disposal
- Minimizing facility siting impacts and citizens' concerns
- Being assured of a long-term source for waste to provide the project

As discussed, developers of recently implemented projects have sorted out their roles and objectives in different ways. This allocation has been a key to determining the position the public sector takes on such projects. Several issues will need particular attention. The ability to secure privately collected waste for the project is essential. If a project is implemented with this waste stream secured under long-term contracts and/or legislatively controlled to assure waste delivery to the implemented energy recovery system, the cost for such a project will certainly be lower than if a project were developed with the waste stream "at risk."

Another critical consideration is whether the project should seek to take advantage of equity funds to lower the debt it will have to repay for the capital facilities required. Several recent projects have been implemented as privately owned projects which offset part of the capital requirements through equity contributions in exchange for tax benefits (leveraged lease or vendor financing structures). Even with public ownership, it is still possible for the project's design, construction, and operation risks to be put on the shoulders of a full-service contractor, as has been done in Auburn, Maine. Additionally, the transfer stations and landfill that are to be part of the system do not have to be owned and operated by the same party/contractor. In fact, some projects have sought to involve local private industry by separating these portions of the project from the energy recovery facility itself.

PROCUREMENT ALTERNATIVES AND RISKS

A crucial decision is the selection of the overall procurement strategy. Three basic procurement approaches exist: conventional A/E, turnkey, and full service.

The conventional approach for procuring any public works project usually involves two separate procurements. First an engineering consultant is employed to participate in the planning and to prepare plans and specifications that serve as the basis for the bids from construction contractors. The same consultant might also be retained to perform such other services as construction inspection or start-up supervision. The second procurement is the awarding of the construction contract.

The turnkey approach assigns to a single party total responsibility for facility design, construction, and start-up. The municipality signs a single contract with a supplier who, upon completion of the contract, turns the facility over to the municipality for operations. There may also be a training and hiring service provided by the turnkey contractor in support of the municipality's operating function.

The full-service approach extends the turnkey concept one step further by including private operation—and possibly private ownership—in the package. (Actually, private operation could be a subsequent contractual step following either a conventional or a turnkey construction contract.) Under a full-service arrangement, the contractor has responsibility for financing, design, construction, operation, and possibly ownership. In essence, the full-service contractor offers the municipality a disposal service, rather than just a physical plant.

Many municipalities have come to regard the conventional A/E approach as unsatisfactory for procuring an energy recovery facility. Using the conventional A/E approach often requires the municipality to assume unfamiliar entrepreneurial and risk-assumption roles. For example, if the plant cannot be mechanically completed or meet performance requirements, the municipality will be required to raise additional capital to complete the plant. The design engineer traditionally does not provide any performance guarantees related to the technology or the service that cannot be provided by an unfinished and/or nonperforming facility. Experience has shown that there has been less focus on the business aspects of a project when the conventional A/E approach has been used. The municipality is also involved, on a day-to-day basis, in design and construction decision making, and later in personnel and maintenance administration. Many public officials are anxious to have private industry assume the maximum amount of responsibility and risk.

The turnkey approach can give the municipality protection from design and initial plant performance risks. Comparing the design changes and plant added costs in Hampton, Virginia (turnkey) with those experienced at the Nashville and Gallatin, Tennessee plants provides examples of the benefits of turnkey versus conventional A/E.

There are cases in which conventional A/E or turnkey is best suited. For example, the A/E or turnkey may be preferable if a municipality (a) has ongoing and well-run utility services (electric, water, or sewer) or a solid waste facility (i.e., transfer station) that can be expanded for resource recovery operations, (b) has an engineering department that can assume major design review and/or construction monitoring functions, (c) intends to operate the plant, (d) has a waste stream already under its control, and (e) has the ability to raise capital and isolate the project from the usual municipal hiring/firing and maintenance/parts budget constraints that typically contain public operations.

The full-service approach has other advantages. It affords the possibility of packaging a complex public/private financial arrangement that offers significant advantages to both parties. In addition, industry generally has proven to be more efficient than government in maximizing the benefits from the sale of recovered products.

Regardless of which procurement approach is chosen, the municipality must issue a Request for Proposals (RFP) that limits proposals to that approach. It is impossible to evaluate a turnkey proposal, for example, against a conventional or a full-service proposal.

The various risks inherent in these projects are outlined in Table 9-3 for each procurement approach. A critical review of each risk should be made and a preference stated as to which party should assume which risk. A completed risk allocation chart for the specific project should be understood locally and presented in any procurement documentation to potential contractors.

PROCUREMENT DOCUMENTATION

The procurement process selected by the project sponsor will delegate many of the risks and responsibilities of the project implementation and operation. The level of risk which is to be assumed by the sponsor can be expressed in its

Table 9–3. Risk Assignment Under Alternative Procurement Approaches.[a]

	RISK ASSUMED BY:		
RISK ELEMENTS	A/E PROCUREMENT	TURNKEY PROCUREMENT	FULL-SERVICE PROCUREMENT
Capital Costs Risks			
Capital costs overruns	O	C	C
Additional capital investment to achieve required operating performance	O	C	C
Additional facility requirements due to new state or federal legislation	O	O	O
Delays in project completion which lead to delays in revenue flow and adverse effect of inflation	O	C	C
Operating and Maintenance Costs Risks			
Facility technical failure	O	C	C
Excessive facility downtime	O	O*	C
Underestimation of facility O&M requirements (labor, materials, etc.)	O	O*	C
Insufficient solid waste stream	O	O/M	M
Significant changes in the solid waste composition	O	O	O/C
Changes in state and federal legislation which affect facility operations	O	O	O
Inadequate facility management	O	O*	C
Underestimation of residue disposal costs	O	O*	C
Tipping Fee Income Risks			
Diversion of waste to other competing facilities	M	M	M
Overestimation of the solid waste stream	O/M	O/M	O/M
Adverse changes in participating community's fiscal condition	O	O	O

(*cont.*)

Table 9–3. (continued)

| | RISK ASSUMED BY: | | |
RISK ELEMENTS	A/E PROCUREMENT	TURNKEY PROCUREMENT	FULL-SERVICE PROCUREMENT
Recovered Energy Income Risks			
Overestimation of technology energy recovery efficiency	O	C	C
Significant change in the solid waste composition	O	O	O/C
Changes in legislation which affect energy production and/or use	O	O	O
Overestimation of solid waste quantities	O	O	M
Significant adverse changes in the energy market financial condition or local commitment	O	O	O/EM
Downward fluctuation in the price of energy	O	O	O
Inability to meet energy market specifications	O	O*	C

a Participants:
   O = owner (assumes owner's ability to assume risk)
   C = contractor
  EM = energy market
   M = participating municipalities
   * = modified turnkey procurements may provide for intermediate or long-term private contractor facility operations which could lead to further risk assumption by the private contractor.

procurement documentation. This section will briefly discuss the Request for Qualifications (RFQ)/Request for Proposals (RFP) procurement method and other approaches such as sole source, RFQ to negotiations, and RFQ to a Basis of Negotiations (BON).

Whether an A/E, turnkey, or full-service procurement method is chosen by the project sponsor, the process usually involves issuing an RFQ and RFP. Table 9–4 shows an example of an RFQ outline. The RFQ requests various background information from respondents including general corporate qualifications, management approach, technical and operating experience, reference plant data, and financial status.

The general corporate information in the RFQ should include the type of company (i.e., corporation, partnership, etc.), history, listing of officers, organizational chart, and project team, if applicable. Management information and approach should demonstrate the respondents' commitment to

Table 9-4. Sample Request for Qualifications (RFQ) Outline.

SECTION NUMBER	TITLE
1.0	Background and Description of Project
2.0	Project Overview
2.1	System Concept
2.2	Waste Stream
2.3	Site Description
2.4	Energy Market
2.5	Environmental Impact
2.6	Contracts
2.7	Financing
3.0	Selection Process
4.0	Schedule
5.0	Information Requested
6.0	Instructions for Preparing Qualifications Statements
7.0	Glossary of Terms and Abbreviations
APPENDICES	
A	Enabling Legislation
B	Procurement Regulations
C	Waste Control Ordinances and Laws
D	Response Forms

energy recovery, especially as it relates to the requested project size. The respondents should provide a listing and description of all previous and current waste-to-energy projects, demonstrating their proven application in similar situations. This should include a list of financial participants, personnel involved, implementation schedules, markets for energy and/or materials, and procurement approaches. The technical and operating experience of the respondent should be fully supported, especially a "reference facility" of similar size and technology. This should describe in detail the equipment, design, technology, operating statistics, and successes/failures of the facility. Finally, the financial status of the respondents is important, especially if a private financing option is to be used. A statement of their net worth; profitability; major financial commitments; performance bond guarantees; equity contributions; contingent liabilities; lines of business and trends in those lines; and indebtness, especially in relation to the reference plant, should be included.

After issuance of an RFQ, the sponsor frequently issues an RFP to selected qualified firms. The RFP outlines the contract structure and details the facility requirements as requested by the sponsor(s). The RFP can set the basis for the negotiations of a final contract. An example of a RFP outline is shown in Table 9–5. The sponsor will then evaluate the proposals based on their responsiveness to the information requested in the RFP. Further information may be solicited in an addendum, if needed. The sponsor(s) then may choose to negotiate with one or more respondents if their proposals are satisfactory.

Table 9–5. Sample RFP Outline
(City of Springfield, MA - Issued 10/7/83)

TABLE OF CONTENTS

Pages

1. PURPOSE AND PROCUREMENT APPROACH
   1.1 Introduction and Project Goals 1–1
       1.1.1 Introduction 1–1
       1.1.2 Project Goals 1–2
   1.2 Responsibilities of the City and Contractor 1–3
       1.2.1 General Responsibilities of the City 1–3
       1.2.2 General Responsibilities of the Contractor 1–4
       1.2.3 Other Requirements of the Contractor 1–4
       1.2.4 General Responsibilities of Other Participating Municipalities . 1–5
       1.2.5 General Responsibilities of Northeast Utilities . . . 1–5
   1.3 Contractor Procurement Schedule 1–6
       1.3.1 Selection of Contractor for Negotiations
             and City's Rights and Options 1–6
       1.3.2 Cost of Negotiations 1–7
       1.3.3 Project Contact 1–7
       1.3.4 Proposal Guarantee 1–7
       1.3.5 Withdrawal from Negotiations 1–7
       1.3.6 Delay in Negotiation Schedule Due to Tardiness
             of Contractor 1–8
   1.4 Content of this RFP. 1–8
   1.5 Summary of City's Project Philosophy 1–9
2. BACKGROUND INFORMATION
   2.1 General 2–1
   2.2 Project History 2–1
   2.3 Existing Solid Waste Disposal 2–2
   2.4 Solid Waste Supply 2–3
   2.5 Energy Market 2–9
   2.6 Residue/By-Pass Waste Landfill 2–10
   2.7 Facility Site 2–11
       2.7.1 Preferred Site - Agawam, Massachusetts 2–11
       2.7.2 Alternate Site - North Centre Industrial Park,
             Springfield, Massachusetts 2–13

Table 9–5. (continued)

TABLE OF CONTENTS (Cont.)

	Pages
2.7.3 Alternate Site - Bircham Bend, Springfield, Massachusetts	2–13
2.8 Local Contacts	2–14
3. FACILITY REQUIREMENTS	
3.1 General	3–1
3.2 Proposer Responsibilities	3–1
3.3 Facility Processing Capacity	3–3
3.4 Redundancy and Reliability	3–4
3.5 Facility Waste Storage Capacity	3–4
3.6 Waste Receiving Requirements	3–5
3.7 Facility Energy Output Capacity	3–5
3.8 General Codes and Standards Applicable to Facility Construction .	3–6
3.9 Facility Layout	3–7
3.9.1 Main Buildings	3–7
3.9.2 Weighing Facilities	3–7
3.9.3 Residue Storage Area	3–7
3.10 Building Architectural and Structural Requirements	3–8
3.10.1 General Requirements	3–8
3.10.2 Selection of Construction Design and Materials . . .	3–8
3.11 Solid Waste Receiving, Handling, and Storage System . . .	3–8
3.12 Residue Handling System	3–9
3.13 Mechanical Equipment General Requirements	3–10
3.14 Refuse-Derived Fuel Processing System	3–10
3.15 Steam Generating System	3–11
3.16 Feedwater Treatment and Supply System	3–13
3.17 Steam Transmission System	3–14
3.18 Turbine Generator System	3–14
3.18.1 General Requirements	3–14
3.18.1.1 Scope of Work	3–14
3.18.1.2 Applicable Standards	3–15
3.18.1.3 General Operating Conditions	3–15
3.18.1.4 General Design Conditions	3–15
3.18.1.5 Equipment Requirements	3–16
3.18.1.5.1 Major Equipment	3–17
3.18.1.5.2 Accessory Equipment	3–17
3.18.2 Technical Requirements	3–18
3.18.2.1 Steam Turbine	3–18
3.18.2.1.1 Type	3–18
3.18.2.1.2 Construction Features . . .	3–19
3.18.2.1.3 Turbine Generator Set Philosophy . .	3–19

(cont.)

Table 9–5. (continued)

TABLE OF CONTENTS (Cont.)

		Pages
3.18.3	Steam Condensing System.	3–19
3.19	Miscellaneous Equipment	3–20
3.19.1	Compressed Air Supply System.	3–20
3.19.2	Feedwater Pumps	3–20
3.20	Aesthetics, Odor and Noise Control	3–20
3.20.1	Aesthetics .	3–20
3.20.2	Odor Control	3–20
3.20.3	Noise Control	3–21
3.21	Facility Control and Monitoring	3–22
3.22	Fire Detection and Control System	3–23
3.23	Environmental Controls.	3–23
3.23.1	Wastewater Effluent .	3–23
3.23.2	Air Quality Criteria .	3–23
3.24	Minimum Performance Standards .	3–24
4.	SCOPE OF WORK	
4.1	Project Development	4–1
4.1.1	Contract for Leasing of the Bondi Island Site	4–1
4.1.2	Contracts With Waste Suppliers .	4–1
4.1.3	Contract for Steam Sales to the City's Wastewater Treatment Plant .	4–1
4.1.4	Contract for Electricity Sales to Northeast Utilities	4–2
4.1.5	Development of a Residue/By-Pass Waste Landfill	4–2
4.1.6	Other Development Activities	4–2
4.2	Financing .	4–2
4.3	Facility Construction	4–3
4.3.1	Project Management	4–3
4.3.1.1	Scheduling of Activities .	4–3
4.3.1.2	Staffing .	4–4
4.3.1.3	Reporting Requirements .	4–4
4.3.1.4	Preparation and Submittal of Facility Operating Plan	4–4
4.3.2	Design Phase.	4–5
4.3.2.1	Requirements for the Design Phase	4–5
4.3.2.2	Submission of Design Review Materials .	4–6
4.3.2.3	Preparation and Processing of Permit Applications	4–6
4.3.3	Construction Phase	4–7
4.3.3.1	Requirements for the Construction Phase	4–7
4.3.3.2	Procurement of Subcontractors	4–7
4.3.3.3	Site Preparation	4–7
4.3.3.4	Preparation and Submittal of Performance Test Plan and Procedure Document	4–7
4.3.3.5	Completion of the Construction Phase .	4–8

Table 9–5. (continued)

TABLE OF CONTENTS (Cont.)

Pages

4.3.4 Start-up Phase 4–8

    4.3.4.1 Requirements for the Start-up Phase 4–8
    4.3.4.2 Preparation and Submittal of Facility Start-up Plan . 4–9
    4.3.4.3 Supply of Waste During Start-up 4–9
    4.3.4.4 Completion of the Start-up Phase 4–10

4.3.5 Performance Test Phase 4–10

    4.3.5.1 Requirements for the Performance Test Phase . . 4–10
    4.3.5.2 Supply of Waste During Performance Test . . . 4–10
    4.3.5.3 Preparation and Submittal of Contractor's
           Performance Test Report 4–11
    4.3.5.4 Review of Performance Test Report 4–11
    4.3.5.5 Completion of the Performance Test Phase . . 4–11

4.4 Facility Operations 4–12

4.4.1 Facility Management and Deliverables 4–12

    4.4.1.1 Scheduling of Facility Activities 4–12
    4.4.1.2 Staffing and Training 4–12
    4.4.1.3 Operations Reporting Requirements 4–12
    4.4.1.4 Annual Facility Performance Review
           and Inspection 4–13

4.4.2 Operation and Maintenance Phase 4–13

    4.4.2.1 Facility Operations 4–14
          4.4.2.1.1 Acceptance of Waste 4–14
          4.4.2.1.2 Removal and Disposal of
               Unprocessibles and Residue . . . 4–15
    4.4.2.2 Facility Maintenance 4–15

5. PROPOSAL INSTRUCTIONS

5.1 General 5–1

5.1.1 Expenses of Proposal Preparation 5–2
5.1.2 Confidential Information 5–2
5.1.3 Number of Proposals 5–3
5.1.4 Cover Letter and Signature Requirements 5–3
5.1.5 Proposal Preparation, Evaluation, and Negotiation Schedule . 5–3
5.1.6 Proposal Guarantee 5–4
5.1.7 Disposal of Proposals 5–4

5.2 Document I: Executive Summary 5–4

5.2.1 General Provisions 5–4
5.2.2 Intended Distribution 5–4

5.3 Document II: Facility Proposal 5–5

5.3.1 Technical Section 5–5

    5.3.1.1 General Provisions 5–5
    5.3.1.2 Facility Technical Description. 5–5
          5.3.1.2.1 Process Flow Diagram 5–5

(cont.)

Table 9–5. (continued)

TABLE OF CONTENTS (Cont.)

	Pages
5.3.1.2.2 Process Control and Instrumentation . .	5–6
5.3.1.2.3 Furnace/Boiler/Air Pollution Control	
System Design	5–6
5.3.1.2.4 Process Mass Balance	5–6
5.3.1.2.5 Process Energy Balance	5–6
5.3.1.2.6 Availability Analysis	5–7
5.3.1.2.7 System Capacity	5–7
5.3.1.3 Site Plan	5–7
5.3.1.4 Environmental Data	5–8
5.3.1.5 Equipment Description and Performance Requirements .	5–9
5.3.2 Management Section	5–10
5.3.2.1 General Provisions	5–10
5.3.2.2 Construction and Procurement	5–10
5.3.2.3 Operation Plan	5–11
5.4 Document III: Business Proposal	5–11
5.4.1 Contract Proposal Section	5–11
5.4.1.1 General	5–11
5.4.1.2 Solid Waste Disposal Service Agreement . .	5–11
5.4.1.2.1 Contractor's Position on Solid Waste	
Disposal Service Agreement	5–12
5.4.1.2.2 Contractor's Position on Electricity	
Purchase Agreement	5–13
5.4.1.2.3 Contractor's Position on Steam Sales	
Agreement	5–13
5.4.1.2.4 Contractor's Position on Residue Disposal	
Agreement	5–13
5.4.1.2.5 Contractor's Position on Site Lease	
Agreement	5–14
5.4.1.2.6 Contractor's Position on Involvement	
With District Heating	5–14
5.4.2 Cost Proposal	5–14
5.4.2.1 Construction Cost	5–14
5.4.2.2 Annual Operating Cost	5–14
5.4.2.3 Financing Plan	5–16
5.4.2.4 Tipping Fee	5–17
5.4.3 Performance Guarantees	5–18
5.5 Document IV: Qualifications	5–20
5.5.1 General	5–20
5.5.1.1 Experience	5–20
5.5.1.2 Technical Reliability	5–20
5.5.1.3 Financial Strength	5–21
5.5.2 Required Information	5–21

Table 9–5. (continued)

TABLE OF CONTENTS (Cont.)

			Pages
	5.5.2.1	General Information	5–21
	5.5.2.2	Experience of Contractor in Solid Waste Energy Recovery	5–22
	5.5.2.3	Technical Reliability of Technology	5–23
	5.5.2.4	Reference Plant	5–23
5.6	Proposal Forms		5–24
	A.	Proposer Information	5–25
	B.	Technical Description of Equipment	5–27
	C.	Environmental	5–37
	D.	Construction Cost Bid	5–41
	E–1	Itemized Capital Cost	5–42
	E–2	Drawdown Schedule	5–45
	F.	Itemized Annual Operating Costs	5–46
	G–1	Financing Assumptions	5–49
	G–2	Sources and Uses of Project Financing	5–50
	H.	Tipping Fee Bid (Proposers' Optional Financing)	5–51
	I.	Performance Guarantees	5–52
	J.	Performance Assurances	5–53
	K.	Default Disposal Fee	5–56
6.	EVALUATION OF PROPOSALS		
6.1	General		6–1
6.2	Evaluation Criteria		6–2
	6.2.1	Primary Criteria	6–2
	6.2.2	Secondary Criteria	6–2

LIST OF APPENDICES

A. City of Springfield Resolution

B. Memoranda of Understanding From Participating Communities

C. Information From Northeast Utilities

D. Site Information: Agawam, Massachusetts

E. Site Information: North Centre Industrial Park, Springfield, Massachusetts

F. Site Information: Bircham Bend, Springfield, Massachusetts

G. Term Sheet: Residue Disposal

H. Term Sheet: Lease of City's Agawam Site to Vendor

I. Term Sheet: Steam Sales to City

J. Executive Summary - Technical and Economic Feasibility Study for District Heating in Springfield

K. City of Springfield Sewer Information

L. Environmental Regulations and Permit Information

M. Draft Solid Waste Disposal Service Agreement

Other procurement processes are available besides the RFQ/RFP approach. The project sponsor may choose to negotiate with a qualified respondent(s) directly after the RFQ process or may develop a Basis of Negotiation (BON) document for discussions with one or more respondents. The sponsor may feel, after the RFQ process, that it is reasonably certain that one or two qualified respondents can provide the waste-to-energy system it desires. If an A/E approach is being used by the sponsor and the design of the facility is complete, the sponsor may begin to negotiate with one or more firms to design and manage the construction of the facility. The operating contractor (if applicable) may be selected well after construction is initiated but preferably earlier. If a turnkey or full-service approach is being used, a Basis of Negotiation (BON) document may be written which outlines the design criteria and contract structure desired by the sponsor as in an RFP. The project sponsor may then negotiate with one or more system vendors to provide these services. The RFQ/BON approach was used for the Baltimore, Maryland Southwest Facility, a large-scale facility, but it is being applied successfully in the procurement of small-scale facilities as well. Both of these processes are successful if the sponsor is sure of the desired technology and system, and is reasonably certain that these can be provided by a particular firm(s). The BON approach permits more open discussion of contract terms before proposals are submitted, thus allowing proposals to be submitted upon prenegotiated contracts.

Finally, a project sponsor may choose, for whatever reason, to negotiate immediately with one firm and not issue any formal request documents. This process is called "sole-source" procurement and requires that the sponsor know the technology and vendor it wants for a project and be legally allowed to take such a step. The sole-source approach expedites the procurement process since there are no RFQ or RFP activities, but it limits the negotiating potential and system alternatives available since there is only one vendor.

The procuring agency might also consider issuing a Request for Developers (RFD) similar to the way some jurisdictions are selecting private groups to develop a shopping center, convention center, hotel/office complex. The RFD approach may be well suited in cases where a private developer has already initiated developing a project, site acquisition is not politically possible, or front-end resources are not accessible for undertaking the project as is usually done. This approach can short-circuit the time to get to a contractor, but not necessarily to get to an operating facility.

It is very important that the sponsor of the project be very familiar with procurement laws in their state and jurisdiction. These may determine the ability of a sponsor to choose the desired procurement process. The sponsor should seek to allow for negotiated procurements with whatever process is chosen.

NEGOTIATIONS

Following the selection of a preferred system vendor, the project sponsor must assemble a negotiating team and prepare for detailed contract discussion. The negotiating team should consist of at least the following project participants: project manager, legal counsel, financial advisor, and management advisor. The coordination and provision of a consistent philosophy and approach by the negotiating team are very important to a contract. Local politicians should be briefed regularly throughout the negotiation process, but they need not participate.

The negotiating team and preferred vendor must both have a clear and well-defined understanding of the contract and risks structure. The contract structure should be remembered throughout the negotiations so that the roles of all participants are clear. Critical "deal" points should be dealt with last; the easily agreed upon issues should be considered first. This will familiarize the negotiating participants with each other and provide a basis from which to continue discussions. It is important to be fair in approach, but always to remember objectives and overall structure. The negotiations will set the tone for the long-term business relationships. The next step after negotiating a contract with the vendor is the financing of the project.

The need for effective coordination and management of energy recovery project procurement and implementation cannot be underestimated. It is essential that the project sponsor and/or management consultant be able to effectively organize the various project participants into a project team which has a consistent philosophy and understanding of the objectives. A project team that is well organized will negotiate a better deal.

The local political climate in the host community is very important to successful project implementation. Local elected officials must be regularly apprised of the project throughout the development process so that they can maintain public support. It is critical that the garbage disposal crisis issue be kept in the foreground at all times. This will accelerate implementation by keeping the focus on the necessity for project development.

It is important for the local jurisdiction to remember, when negotiating a project, that it has the power to make the project happen and thus carries bargaining strength. Contractors rely upon the local jurisdiction to implement the project. If a project team bargains too hard, this will increase the project cost; thus it is important to be consistent and fair. This approach will achieve the best negotiated contracts for all participants.

PUBLIC INVOLVEMENT

During the course of the project's development, the public will need to be informed, educated, and involved with the project development team and, likely, with elected officials. The public's reaction to a project has often been negative, attracting the media's attention and resulting in negative headlines that will work against building public support for a new solid waste management system. Community groups often organize to fight a site selected near their homes. Issues such as truck traffic, noise, rodent infestation, odor, and air pollution are raised as significant impacts. Particularly troublesome can be dealing with much misinformation that is generated (Chapter 8 deals with these issues in more detail).

The point is that the project manager needs to undertake a sincere and honest communication effort toward the media and citizens. Communication tools can include publishing monthly newsletters; providing speakers for social, business, and community organizations; issuing press releases and short articles for local publication; taking media and community representatives to similar plants that are operational; having displays for fairs and exhibits; distributing special informational brochures to residences and schools to explain the need and the solution; and holding public forums/hearings in conjunction with key project milestones.

Doing this may not always work, but it usually helps give the project a positive image. These efforts do not have to cost an exorbitant amount to put together, and new projects can draw on the excellent materials prepared in support of projects in other communities.

10
THREE SMALL-SCALE
WASTE-TO-ENERGY
SYSTEM CASE STUDIES

The three case studies selected for inclusion in this chapter illustrate three distinct approaches to small-scale waste-to-energy processing. The facility located in Auburn, Maine is a modular, starved air system (Consumat), with a design capacity of 200 TPD (tons per day). In Madison, Wisconsin, RDF is produced and transferred off-site where it is co-fired with pulverized coal in two Babcock and Wilcox suspension boilers. The third case study, Pittsfield, Massachusetts, describes a small-scale modular excess air mass burn system designed by Vicon Recovery Associates to process 240 TPD.

The Madison facility is a joint venture between the City of Madison and the local utility, Madison Gas and Electric. The RDF/suspension boilers produce electricity for MG&E's system. Both Auburn and Pittsfield are towns which took the lead in addressing their area's solid waste disposal problems, and both accept waste from surrounding communities. The two New England facilities produce steam to meet the specifications of a single industrial customer. The Pittsfield facility supplies steam to Crane and Company, a major paper manufacturer. In Auburn, the energy customer, Pioneer Plastics, joined the planning team (Auburn Solid Waste Review Committee) to implement the system.

AUBURN, MAINE

Introduction

The Auburn, Maine Resource Recovery Facility is located south of the downtown area adjacent to Pioneer Plastics, the energy purchaser. Auburn is a community of approximately 24,000, situated on the western shore of the Androscoggin River 32 miles north of Portland and the Maine seacoast. The City's river location attracted shoe and textile manufacturers. Many of these are still located across the river in Lewiston. Auburn, however, has seen a transition in its employment base to more diversified industry, with Pioneer Plastics (a division of Libbey-Owens Ford), General Electric, and Tampax,

Inc. as its primary employers. Pioneer Plastics is Auburn's largest employer with over 700 employees.

The Auburn facility is a modular controlled air waste-to-energy facility with a design capacity of 200 TPD. Waste is supplied from Auburn and more than a dozen smaller communities. Manufacturing wastes from Pioneer Plastics and oil spill cleanup wastes from the Maine Department of Environmental Protection are also processed at the facility.

Start-up operations at the facility began in April 1981, and the plant presently processes an average of 160 TPD. The facility is owned by the City of Auburn and operated by Consumat Systems, Inc. of Richmond, Virginia, the equipment supplier.

Background

In 1974, the Maine Department of Environmental Protection (DEP) adopted regulations affecting the operation of land disposal sites in the state. These regulations required open dumps such as the one then operated by the Auburn Public Works Department to close or be converted to sanitary landfills. In addition, Auburn's dump was located near Lake Auburn, the City's water supply, providing further incentive to discontinue landfilling at this site.

In response, Auburn and a group of surrounding communities, which were also faced with conforming to the regulations, formed a Regional Solid Waste Committee (RSWC) to explore alternative disposal options. Charter members of the RSWC included six communities ranging in size from 1,000 to 24,000 population. A consultant hired to evaluate the area's solid waste disposal alternatives identified two options in May 1975: a regional landfill in Auburn and an energy recovery program based on a potential contractual arrangement with an Auburn industry (Pioneer Plastics).

In March 1977, the Auburn Public Works Department prepared a report to aid City decision makers in developing an implementation strategy. The report dealt with a specific definition of Auburn's solid waste problem and development of a strategy to achieve an energy recovery project or, if necessary, a sanitary landfill facility. In conjunction with the report, a limited weighing program was undertaken. This program revealed the following:

- The average waste quantity being disposed of in Auburn's landfill was approximately 100 TPD (commercial, industrial, and residential).
- Over 20 percent of this waste, by weight, originated from Pioneer Plastics' manufacturing of decorative laminates.
- The Pioneer waste material consisted of two distinct waste categories:

(1) broken laminates and regular mill waste, and (2) a fine sawdust-textured material which results from the laminate manufacturing process.

In early spring 1977, a City in-house review committee was formed as an implementation task force. Its composition reflected an appreciation of the complex mix of technical and broad community-interest components important to Auburn's eventual decision. This committee included the City Manager, the City Engineer/Public Works Director, the City Finance Officer, the City Purchasing Agent, and two members of the City Council.

To help the committee thoroughly investigate the available recovery technologies, the City applied for and received EPA technical assistance. This aid focused on the technical and institutional aspects of the system's implementation. A strategy option report was prepared dealing with the pros and cons of private/public ownership and operation, the risks inherent to these various options, and the methodology of risk avoidance, risk sharing, or risk assumption. From this analysis, a concept of municipal ownership of a guaranteed system with limited private operation evolved.

The specific planning issues that followed centered around four areas:

- Energy market
- Technology and contractor selection
- Residue disposal and landfill backup
- Regional community waste supply

It was decided, at this point, that the City would finance the project with existing fund reserves and general obligation issues. A Request for Proposals meeting Auburn's identified needs was developed. The procurement process utilized a modified turnkey approach. To ensure the selected contractor's intimate involvement in the project, the RFP was written so that the contractor would be required to provide:

- Complete architectural and engineering design in accord with detailed performance parameters
- All steps needed for the purchase of the necessary structure and equipment
- All construction services for the approved project design on the City's site adjacent to the steam market, Pioneer Plastics
- All services necessary for plant start-up and trial operation of the facility in its entirety, and to establish operability of all component systems and equipment

- All services necessary for plant acceptance in conformance with the performance requirements and guarantees
- Operating, maintenance, and output control services for a period of three years from the initial date of project acceptance

Proposals were requested for two basic system modes: System Mode 1 would have the capability to process 100 TPD; while System Mode 2 would have a capacity of 220 TPD. Potential contractors were required to provide a proposal for each mode. Optional proposals were requested on ferrous metal recovery and sewage sludge disposal.

Contractors who submitted proposals were required to provide system guarantees for: (1) steam quantity and quality, (2) air emissions, (3) solid waste volume reduction, and (4) supplementary fuel consumption.

The RFP was issued on December 1, 1977. On February 28, 1978, proposals were received and six were accepted for evaluation. These proposals represented a variety of different technologies, with capital cost for a Mode 1 system ranging from two million to over six million dollars.

Three were determined to be responsive: Consumat Systems, Inc.; Waste Management, Inc.; and Envirotech Corp. The Consumat proposal was selected as the best submitted. The 100 TPD System Mode 1 was selected with three 50 TPD units to be installed, assuring availability of 100 TPD processing capacity. In addition, the facility design was to allow room for a fourth unit. This fourth unit was purchased with monies from the Maine Department of Environmental Protection and was installed while the facility was still under construction. A modified turnkey contract with Consumat was signed in October 1979. Under the terms of this agreement, Consumat would operate the facility for three years. The City of Auburn as owner would pay Consumat an operating fee for waste processed.

The contractor (Consumat) guaranteed that the process and equipment parameters provided could be consistently met during normal operations, if the equipment was operated in accordance with the manufacturer's instructions. Specifically, the following process and equipment parameters were addressed:

- Air emissions (0.08 grain per standard cubic foot corrected to 12 percent CO_2)
- Auxiliary fuel consumption (500,000 Btu per ton processed)
- Process capability (4,200 pounds of waste per hour)
- Energy (steam) production (4,800 pounds of steam per ton of waste processed)
- Ash residue (less than 5 percent by volume)

Auburn recognized early that no matter what solid waste management system was implemented, a landfill would still be necessary. However, one of the reasons for Auburn's consideration of an energy recovery system was the depletion of capacity in its existing landfill. An assessment was made of remaining capacity to serve the City during the time necessary to implement the energy recovery system. It was decided to design an interim landfill in proximity to the existing facility. This site was designed to serve the City for approximately two years.

Simultaneously with the effort to develop an interim landfill, it was decided also to seek approval on an ash residue landfill. One of the sites originally evaluated for development as a sanitary landfill was felt to have potential for this use. Although within 5 miles of the proposed processing facility, the site had been rejected as a potential regional landfill because of its limited capacity. However, the volume and weight reduction anticipated from the energy recovery system allowed the site to be considered for a bypass/ash disposal facility. A long-term lease was negotiated with the Maine Turnpike Authority (the existing land owner) for this use.

The plans for both facilities were submitted to local boards and to the Maine Department of Environmental Protection, and approval was received for both facilities. Although initial studies considered a regional approach, Auburn had independently pursued implementation of the system. In order to evaluate the economies of scale, proposers were asked to submit two cost proposals: one for a system serving Auburn alone and one providing capacity for surrounding communities.

The proposals received indicated that a definite economy of scale existed in the technologies evaluated. The modular nature of the selected technology easily allowed for different capacities based on the number of modules installed.

Auburn then proceeded to negotiate waste supply contracts with surrounding communities. Fourteen communities now have long-term (20 year) contracts with the City. The City presently supplies approximately 100 TPD, with the contract communities delivering 60–80 TPD. The project chronology is summarized in Table 10–1.

Technology

The energy recovery facility is situated on Goldwaithe Road adjacent to the Pioneer Plastics plant. As the public is not allowed to deposit waste in the main facility building, a private residents' disposal area is located in front of the plant with four bins in a semienclosed structure. Incoming commercial waste (or municipally collected waste) is weighed at a truck scale, then

Table 10–1. Auburn, Maine Modular Facility Chronology.

1974	State stiffens landfill requirements Regional Solid Waste Committee formed
1975	Feasibility study published
1977	EPA technical assistance awarded RFP issued (December)
1978	Contractor selected (July)
1979	City signs contracts with Consumat and Pioneer Design/construction begins (October)
1981	Start-up/shakedown/operation (April)

(Note: Consumat proposed a 14 month schedule from ground breaking through acceptance testing)

discharged onto a tipping floor. Control of the truck scale is accomplished from remote readout at the facility office.

The plant consists of four 50 TPD Consumat modular incinerators with two shared boilers. The facility was initially to have an installed capacity of 150 TPD, with space available for the addition of a fourth 50 TPD module. This additional module was added through a contract with the Maine Department of Environmental Protection for the disposal of oil spill debris. The State furnished the capital costs for the fourth furnace. The final agreement between Auburn and the State of Maine provided that, although the City was given a certain time requirement for processing oil spill debris, the capacity of the fourth unit could be utilized by the City under normal conditions to process other solid waste for energy production.

Raw waste from the tipping floor is fed into the charging hopper of one of four Consumat units. The charging rate is determined by an indicator panel located above each feed hopper. A hydraulic ram automatically injects the waste into the incinerator primary chamber. Hydraulic rams move the burning materials through the chamber along stepped hearths. The combustion environment in the primary chamber, maintained at substoichiometric or starved air conditions, limits the combustion in this chamber. Gases generated in the primary chamber are fed to an upper "pollution control" chamber (secondary combustion chamber) to be mixed with additional combustion air and are combusted under controlled conditions for maximum energy recovery. Gases exiting from the recovery equipment are discharged through a stack. Hot gases may also be vented through a bypass stack when energy recovery is not required. The ash material from the combustion process is ejected from the primary chamber into an ash discharge chute leading to a common quench trough. Ash is then conveyed to an ash removal bin for transport to the residue landfill site.

The generated steam at 295 psig with 60–100°F superheat[1] is piped approximately 100 feet to the Pioneer Plastics plant. Fifty percent by volume of the steam flow is returned as condensate. The facility has been in operation since April 1981. By July 1981, the facility was accepting waste from 14 communities and selling an average of 30,000 pounds of steam per hour to Pioneer.

During the first year of operations, it became evident that due to load management difficulty and a drop in Pioneer's steam demand because of the general economy, a portion of the available steam flow from the facility would not be sold. The ability to "dump" a portion of the steam flow was retrofitted into the system.

Additionally, the facility did not pass its initial performance tests. The primary test parameter not met was air quality. Subsequent investigations after the test period indicated that the facility's baghouse had been bypassed and, therefore, did not function during the tests. With correction of this problem, the next series of tests were passed in August 1982.

Additionally, the City of Auburn is now going to a seven day per week operating schedule due to Pioneer's energy demand and available waste supply. Modifications to the processing system allow alternate isolation of half of the total system (two incinerators, one boiler) for weekend operation and maintenance.

Markets

The 1975 study of potential users of energy indicated that Pioneer Plastics was the only viable energy customer within the RSWC area (two other potential customers were evaluated and rejected). Consequently, Pioneer Plastics representives were invited to join in forming the Auburn Solid Waste Review Committee to implement the system. Pioneer expressed interest in participating for the following reasons:

- A lower cost might be experienced in the development of the energy (steam) required by Pioneer for both heating and processing.
- The fuel feedstock (solid waste) for the facility could provide a captive source of energy, while fossil fuels might be curtailed or regulated in the future.
- The quantity of solid waste from Pioneer (20 percent of the total of Auburn's waste) had to be hauled 10 miles (round trip) to Auburn's existing landfill.

---

1. In this context, "degrees of superheat" refers to the difference in temperature between steam generated for superheat application and saturated steam at the same pressure.

- Pioneer's position as one of Auburn's major taxpayers generated an interest in Auburn municipal matters.

Pioneer required superheated steam at 500–600°F and a maximum pressure of 285 psig. To meet Pioneer's steam demand, it was determined that an average of 50,000 pounds of steam per hour was needed. Much of this steam was used for processing; therefore, the demand would be moderately constant throughout the year.

Negotiations with Pioneer secured an agreement which would require Pioneer to assume its share of the inherent "business risk" of energy recovery. Major points of this agreement were as follows:

- The purchase price of steam generated within the facility will be adjusted with changes in fuel oil costs.
- Auburn will guarantee to produce a minimum of 15,000 pounds of steam per hour at the previously stated quality.
- Pioneer will guarantee the purchase of 15,000 pounds of steam per hour (360,000 pounds over a 24 hour period) and 93,000,000 pounds per year for the 20 year duration of the contract.
- Pioneer will deed 5 acres of land for the facility to the City as a part of its commitment to the project.
- Pioneer will produce the rest of its steam requirements utilizing its existing boiler system. However, the energy recovery facility will provide baseload steam.
- Condensate will be returned to the energy recovery facility with a credit given for returned condensate energy.
- Should Libbey-Owens Ford (LOF) close its Auburn plant, LOF will pay all remaining principal and interest on the energy recovery facility, and total facility operating costs for two years after its closing. Principal is calculated at $170,000 per year and interest at almost $200,000 during the second year, declining for subsequent contract years.

The City's responsibilities include delivery of the specified steam minimums stated, during an operating period to run from 11 P.M. on Sunday through 11 P.M. on Friday, 51 weeks per year, for 20 years. Additionally, the City will accept Pioneer's solid waste material at no cost. The contract was signed in October 1979.

Economics

The financial structure of the Auburn project is a function of the planning and negotiations between participants which occurred in developing the project. The relationships and contracts between the primary project participants were developed to provide adequate and equitable business risk sharing among the participants. The City's overall objective can be illustrated as:

Ownership Costs + Operating Costs − Revenues = Auburn's Costs

Ownership costs include the debt service payments on the facility as well as all costs for planning and developing the facility and the required ash residue landfill. Operating costs include all charges for operating the waste-to-energy facility and ash residue facility. Revenues include tipping fees from the 14 other participating communities and energy (steam) sales revenues obtained from Pioneer Plastics.

The procurement approach adopted by Auburn was a modified turnkey, with the City contracting with Consumat to operate this facility for three years. At the end of this period, the City may elect (1) to operate the facility, (2) to extend the contract with Consumat, or (3) to bring in a third party operator.

To expedite development, the City entered into a design construction agreement with Consumat. City costs for facility construction and development included such items as access road reconstruction, legal and implementation consultants, and a "clerk of the works." The City was also responsible for development of the ash and bypass landfill. Although design and development of the landfill were performed by City personnel, the approximate real value of development was $100,000. The annual lease payment of $100 is made to the Maine Turnpike Authority.

Monies for the facility development came from a number of sources, including:

General Obligation Bond Proceeds	$3,400,000
Maine Department of Environmental Protection Grant	564,000
U.S. Department of Energy Grant	300,000
City of Auburn General Revenue Sharing Funds	25,000
Interest Earned on G.O. Bond Proceeds through Aug. 31, 1981	427,589
	$4,716,589

The general obligation bond funds, resulting from a joint issue including the construction of a school building, were provided through the State of Maine Bond Bank. Construction funds were placed in reserve to provide periodic payments to the project contractor on regularly scheduled requisitions based on construction completion. The bond amortization period is 20 years. The actual project costs, including development of the ash disposal facility, totaled $4,902,359. A breakdown of these costs is presented in Table 10-2.

Operating Costs.   In accordance with the terms of the initial contract between the City of Auburn and Consumat Systems, Inc., the facility contractor was to operate the Auburn waste-to-energy facility for a minimum of three years. During the operating year, Consumat was to be paid $429,555 for processing an initial base solid waste quantity of 26,000 tons. To process solid waste in excess of the base quantity, the contractor was to be paid $8.04 per ton.

For each subsequent year of the operation according to the City of Auburn/contractor agreement, the fee (base and excess tonnage fee) will be adjusted by the Consumer Price Index—All Urban Consumers (CPI-U) published by the Bureau of Labor Statistics, U.S. Department of Labor, for the greater Boston area. Through the contract, Consumat Systems, Inc. must operate and maintain the facility in a prescribed and workmanlike manner, meeting guaranteed performance standards.

However, as the plant began operations, the benefits of increasing the operating period from five to seven days per week became obvious, and certain changes in the fee structure were negotiated accordingly.

Table 10–2. Auburn, Maine Modular Facility Project Capital Costs (1979 dollars).

Facility Construction/Development	
Facility Design (CSI)	277,586
Facility Construction (CSI)	3,974,789
Implementation Planning Consultant	25,000
Codisposal Feasibility Consultant	30,000
Access Road Reconstruction (by City of Auburn)	18,731
Bond and Legal	4,711
Survey	2,160
Test Borings (for EFP development and ash disposal site development)	10,700
Clerk of Works (construction monitoring)	16,623
Interest Expense (construction period)	336,136
Debt Service (construction period)	170,000
Ash Disposal Facility Construction/Development	
Geologic Consultant	16,300
Test Wells	8,397
Materials	11,226
Total	4,902,359

Revenues.   Major revenues to the project are derived from tipping fees received from the communities that are transporting solid wastes to the facility and from the sale of energy (steam) to Pioneer Plastics. Revenues are received by the City of Auburn with payments made on the debt service and to Consumat under terms of the operating contract.

Municipalities interested in bringing waste to the City of Auburn facility were given a variety of participation options. Communities were allowed to consider a 5, 10, or 20 year term agreement with the City. Of the 160 TPD of municipal solid waste (MSW) processed at the facility, approximately 50 percent is received from other participating communities. All communities under contract have decided on 20 year terms and are split between those that will pay a set fee for the contract term and those that will pay a fee which fluctuates with Auburn's cost for the preceding year. Each community is directly responsible for transporting its own waste to Auburn.

Significant emphasis was given to the revenues derived from steam sales. In the agreement between Pioneer Plastics and the City, two alternative steam price formulas are utilized. The actual monthly billing is based on the steam formula which determines the lowest charge. In the first alternative formula, Pioneer's steam charge is based on its existing cost of No. 6 fuel oil. The second formula bases the price of steam on the Consumer Price Index.

Under the first alternative, the initial commodity charge "base price" is calculated as follows:

$12.18/bbl (oil) ÷ 6.3 million Btu/bbl = $1.93/million Btu (oil)

$1.93/million Btu (oil) ÷ boiler efficiency (%) = $/million Btu (steam)

$/million Btu (steam) × 1.235 million Btu/thousand lb (steam) = $2.29/thousand lb (steam). The boiler efficiency is assumed to be 84.4% and will be adjusted if Pioneer's boiler efficiency is proven to have changed. The current monthly charge is then calculated as follows:

$$\text{Current Monthly Charge (\$/thousand lb steam)} = 2.83 \left[1 + .75 \left(\frac{\text{Posted Price Current Month}}{\text{Posted Price Base Month}} - \frac{\text{Posted Price Base Month}}{} \right) \right]$$

In this calculation, the "posted price current month" is Pioneer's current monthly invoiced cost for No. 6 fuel oil delivered to Pioneer's plant in Auburn. The "posted price base month" is set at $12.18 per barrel. Based on Pioneer's existing boiler efficiency, the base price of steam is $2.83 for calculation of the current monthly charge. In the unlikely event that the posted price for the current month is lower than that for the base month, the monthly charge is calculated as follows:

Posted price current month (for month in which posted price is below posted price base month) ÷ 6.3 million Btu/bbl = $/million Btu (oil)

$/million Btu (oil) ÷ boiler efficiency (%) = $/million Btu (steam)

$/million Btu (steam) × 1.235 million Btu/thousand lb (steam) = $/thousand lb (steam)

Under the second alternative, the steam price is set at $6.36 per thousand pounds of steam at the time of first commercial steam sale, adjusted (either up or down) for changes in the Bureau of Labor Statistics Consumer Price Index. The base price of $6.36 per 1,000 is divided by the Index number for the month preceding the month in which the term of the contract commences, and the resulting quotient is multiplied by the Index number for the month preceding the month with respect to which the price calculation is being made.

Through the original energy purchase agreement, Pioneer was to purchase a minimum of 15,000 pounds of steam per hour for a period of 260 days a year (24 hours per day). On a yearly basis, an average throughput of 160 TPD would provide processing for (260 × 160) 41,600 tons of solid waste. Based on the manufacturer's warranty of 4,800 pounds of steam per ton of refuse, the following steam production revenues had been anticipated, with the assumption that the steam value through the formula remained the same and that all steam would be sold:

$$\frac{41,600 \text{ tons}}{\text{year}} \times \frac{4,800 \text{ lb}}{\text{ton}} \times \frac{\$5,952}{1000 \text{ lb}} = \$1,188,495$$

There is a provision contained within the City of Auburn/Consumat Systems, Inc. agreement whereby Consumat Systems, Inc. will receive 25 percent of any revenue derived from the sale of steam above the 4,800 pound per ton level. If, for example, Consumat Systems, Inc. is able to achieve an average steam production rate of 5,300 pounds per ton and all steam can be sold, Consumat will receive 25 percent of the total revenue for the additional 500 pounds per ton of steam which is produced. The City of Auburn was very willing to share any potential excess revenue with the facility contractor since this was not considered in the decision making analysis of economics. As previously mentioned, soon after start-up the plant went to a seven day per week schedule, and the operating economics for April 1983 through March 1984 are shown in Table 10-3.

Table 10–3. Auburn, Maine Modular Facility O&M Revenues/Expenses (thousand dollars).[a]

I.	*Revenues*	
	Steam	715
	Tipping Fees (except Auburn)	320
	Subtotal, Annual ($/ton)	1,035 (18.96)[b]
II.	*Expenses*	
	Base Management Fee (to Consumat)	714
	Ash Residue Facility	114
	Miscellaneous	4
	Subtotal, Annual ($/ton)	832
	Debt Service (principal and interest)	405
	Subtotal, Annual ($/ton)	1,237 (22.65)
III.	*Net Expense (to Auburn)*	
	Annual ($/ton)	202 (7.73)[b]

[a] Period from April 1983 to March 1984.
[b] Based on 54,600 tons processed, of which the City delivered an estimated 26,000 TPY.

Project Evaluation

In developing the project, the City of Auburn was seeking to provide a cost-effective, environmentally sound, socially acceptable method of solid waste disposal. The project has achieved all of these goals. The system as it exists is cost effective when compared to costs for a properly operated sanitary landfill. Present City costs are approximately $8 per ton. Additionally, the public/social perception of the project is very favorable, with only minor complaints received concerning normal facility operations. Also, the facility has been reliable in processing *all* processible waste brought to it during the initial three year operating period, with no unscheduled waste bypass.

However, in choosing to develop a waste-to-energy facility, the City of Auburn has had to adopt a business/production attitude concerning the project. Accordingly, the City has made significant efforts to optimize project factors. The major shortfall has been Pioneer Plastics' inability to continually purchase all available steam. The general condition of the economy has resulted in a reduction in Pioneer's steam needs, with the result that only 75 percent of the available refuse-derived steam flow is purchased by Pioneer. A steam use analysis of facility steam production and Pioneer steam demand curves is in progress. Beyond reviewing Pioneer's use of the available steam, this analysis will evaluate the potential for cogeneration of electricity.

A summary of this project is presented in Table 10–4, and 1983 performance statistics are shown in Figure 10–1.

Table 10–4. Auburn, Maine Modular Facility Case Study Summary.

A. *General Characteristics:*
1. Location — Goldwaithe Road, Auburn, Maine
2. Owner — City of Auburn
3. Procurement Approach — Modified turnkey
4. Design Capacity — 200 TPD
5. Process — Modular (4 units @ 50 TPD)
6. Major Equipment Vendors — Consumat—furnaces and boilers
   Michigan Boiler—baghouse
7. Operator — Consumat initial $3\frac{1}{2}$ years; now City operation
8. Start-up Date — April 1981
9. Current Status — Full-scale commercial operation
10. Energy Market — Pioneer Plastics
    A division of Libbey-Owens Ford
11. Financing Mechanism — G.O. bond with state grant

B. *Economics*

1. Capital Cost ($1,000)

G.O. Bonds	3,400
State Grant	564
Miscellaneous	938
Total	4,902 (1979)

2. Operating Cost (April 1983–March 1984)

	THOUSAND DOLLARS	DOLLARS PER TON[b]
O&M	832	15.24
Debt Service	405	7.42
Subtotal	1,237	22.66

3. Revenues (April 1983–March 1984)

Energy	715	13.10
Tip Fees (except Auburn)	320	5.86
Subtotal	1,035	18.96

4. "Actual" Net Tipping Fee Cost (average)[b]

	522	9.56

C. *Performance*
1. Throughput vs Capacity — 93 TPD average throughput (46% capacity)[a]
2. Major Technical Problems
   During Start-up — Air pollution control equipment improperly installed
   During Commercial Operation — Energy market's demand lower than projected and difficulty following load
3. Other Major Problems — Waste control achieved through series of individually negotiated contracts
4. Future Changes — Seven day per week operation; evaluation of cogeneration potential

[a] Based on design capacity of 200 TPD.
[b] Based on 54,600 tons processed (est. by City Manager).

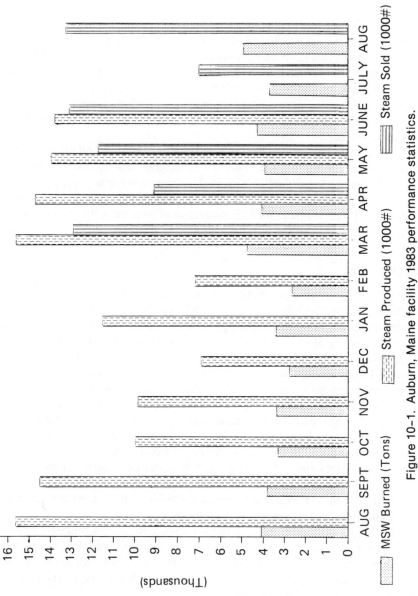

Figure 10–1. Auburn, Maine facility 1983 performance statistics.

231

MADISON, WISCONSIN

Introduction

The City of Madison and the Madison Gas and Electric Company (MG&E) are cooperating in a joint municipal solid waste-to-energy system, which involves the production and use of refuse derived fuel (RDF). The RDF is produced by the City and co-fired with pulverized coal by MG&E in two Babcock and Wilcox suspension boilers, each of which drives a 50 megawatt electrical generator. RDF from the facility is also fired in boilers at the Oscar Mayer Company.

The Madison RDF processing system is included in this collection of case studies because it is one of the few small-scale RDF facilities that is selling its product on a regular and reliable basis. This system is not without problems, especially when viewed from the standpoint of economic viability. However, at this point, the Madison facility appears to offer a workable blueprint for other small-scale RDF facilities to follow.

Background

The City of Madison's experience with alternatives to conventional landfill disposal predates the RDF system. In 1967, the City began a test program of shredding solid waste prior to disposal as a means of conserving landfill space. This program was funded by a demonstration grant from the federal government's solid waste program, which was part of the Department of Health, Education and Welfare at that time. The project was later expanded to full-scale operation, handling all of the City's solid waste.

In the early 1970s, the City began to look at energy recovery as a means of further reducing the demand for landfill space. A variety of factors led to the decision to employ RDF as the energy recovery approach. Perhaps the most important factors were the City's shredding experience and the market requirements. At the time, shredding was considered an integral component of any RDF system and the City had extensive experience with this operation. Furthermore, MG&E expressed interest in the use of RDF as a supplemental fuel in its coal-fired power plant.

The shredding-landfill operation continued until it was replaced by the RDF processing system in 1978. The shredding plant was then closed for removal of the existing equipment. Some of the equipment (conveyors, compactor) was salvaged for use in the new system. A chronology of the project development is outlined in Table 10-5.

The organization of the Madison energy recovery project was shaped by the City's shredding-landfill experience. On the surface, the relationships

Table 10-5. Madison, Wisconsin RDF Facility Project Chronology.

DATE	EVENT
1967	City begins test of shredding MSW prior to landfill
1972	Ferrous metal recovery subsystem installed at shredding facility
1973	Initial discussion between City and MG&E about use of RDF
1974	Engineering feasibility study of RDF use in MG&E boilers
1976	City feasibility study
	City begins negotiations with MG&E (April)
1977	Agreement between City and MG&E to use RDF. Madison Solid Waste Fuel Co. agrees to supply processing system (November).
1978	DNR grants final approval for facility (April; process began 11 months earlier). Construction begins on modifications necessary for RDF processing facility (April). Shredding operation ceased.
1979	Facility processing of MSW
1980	MG&E begins burning RDF
	Testing begins at Oscar Mayer
	Secondary shredder accepted (November)
1982	Oscar Mayer completes modifications to its existing air pollution control system. DNR approves full-time burning at Oscar Mayer (June).

among the participants appear to be typical. The City operates all of the RDF processing system and owns most of it. The City sells the RDF product directly to MG&E, but recovered materials are sold through an intermediary. Waste supply is controlled by the City through public collection. However, the basis for these relationships is related to the City's previous shredding project.

An important facet of the shredding project was the addition of ferrous metal recovery in 1972. This subsystem was installed by Continental Can Company (CCC) at its own expense. In the mid-1970s, CCC sold its system concepts to Combustion Engineering (CE), which still owns the rights to the Madison system. In contrast to CCC's active participation in the project, CE has never been directly involved with the City.

As a result of their early involvement in the waste processing field and desire to keep active in it, several CCC engineers left the company and, thereafter, these individuals played important roles in the development of the Madison project. One of them purchased the flail mill and the separator trommel from CCC. He formed Madison Magnetic Operations (MMO) to continue ferrous separation for Madison.

MMO, under its agreement with the City, owns the magnetic separation equipment and paid for its installation. The City's plant crews operate and maintain the equipment and bill MMO; the City also receives 10 percent of the after-cost revenues from the sale of the metal. Madison Magnetic currently markets the metal to a steel mill in Chicago. Another company, Madison Solid Waste Fuel Company (MSWF), owns the "separation unit" (the trommel). As with the magnetic separator, the company furnished the equipment and paid for its installation; the City operates and maintains the unit and bills the company. The trommel was installed under a ten year performance contract; it is guaranteed to process 50 tons per hour (TPH) of input, with a minimum RDF output of 30 TPH at a maximum of 15 percent ash content. The City pays MSWF an annual fee, plus a variable per-ton fee based on throughput.

Another unique aspect of the project is that the City, through its engineering department, designed the complete system. Specialized design assistance was provided by another former CCC engineer. The City also served as prime construction contractor. The fact that the City was willing and able to perform these tasks is directly related to the experience gained from the shredding operation. City employees at the shredding plant who were transferred to the RDF facility were taught all phases of the RDF processing system, and many even helped to construct it. The City's operating crews also attended a three week training course conducted at a local vocational-technical school. This contributed to a shorter shakedown period and more confident, capable operators.

Assuring the waste supply is straightforward since most of the waste entering the system comes from City operated collection. Two small neighboring towns also contribute solid waste to the system. The system is currently averaging around 270 TPD, but it is capable of processing a much higher throughput. However, surrounding communities are reluctant to participate in the system because the tipping fee is higher than at currently available landfills.

Technology

Processing MSW to produce RDF and the combustion of RDF are the two stages of the Madison system, as is typical with any RDF project. RDF is produced at the City operated processing facility and then trucked to an MG&E electric power plant.

RDF Processing.   The MSW is delivered to the processing plant where it is dumped onto the tipping floor. Small front-end loaders are used to push the refuse into a 2 foot deep pit that has a metal pan conveyor in the bottom.

The first processing point in the system is the primary shredder, which is actually a flail mill. The purpose of this initial shredding is primarily to open refuse-filled plastic bags. The average MSW particle size exiting the flail mill is 8 inches. Madison has addressed the explosion problem, typical of shredder operations, by isolating the flail mill with reinforced concrete and masonry walls. In addition, blowout panels have been installed in the roof.

Refuse leaving the flail mill is conveyed beneath a drum electromagnet, which removes the ferrous metals from the waste stream. Separated ferrous metals are upgraded through densification and removal of light contaminants by an air scrubber prior to shipment to a customer.

The refuse is then conveyed to a trommel for the next stage in the processing line. The majority of the remaining noncombustible materials, such as glass and grit, are separated from the combustible fraction in the trommel. These noncombustibles are taken to a stationary compactor for densifying and storage before landfill disposal.

Combustible materials exiting the trommel are conveyed to the secondary shredder. The objective at this point is to produce a material suitable for combustion in the MG&E boilers, which require $\frac{3}{4}$ inch particle size.

Immediately upon discharge from the secondary shredder, the material enters an air chamber. This processing step functions both as a pneumatic transport system and as an air classifier. Heavy materials (e.g., wood, wet garbage) drop through the air stream to a stationary compactor prior to landfill disposal. The material remaining in the air chamber is blown into a cyclone and de-entrained; then it is conveyed to a transport trailer for transfer to the MG&E power plant approximately 3 miles away.

RDF Combustion.   RDF is unloaded from a trailer into one of two storage bins (about 110 cubic yard capacity) at the MG&E receiving facility. These bins were designed to be small because storage of more than a few hours was not desired.

Each bin is equipped with an auger that feeds the RDF onto a conveyor. The RDF then is taken to flutter, air lock feeders and is pneumatically injected into the boilers. The RDF is fed into each boiler from two refuse nozzles along with six coal nozzles. Because of a problem with burning refuse in the dry ash pit, drop grates were installed just above the pits to allow for combustion of the RDF that fails to burn in suspension. To aid combustion, fans were installed for overfire and underfire air.

RDF supplies about 15 percent of the heating value of the normal fuel, which is pulverized coal. Since this MG&E plant is used for peaking power, the company requires a 60 percent load on the boilers before using RDF. Consequently, RDF is used only during high-load periods, which amount to about ten hours per day, five days a week.

Markets

In 1977, the City of Madison and MG&E signed a ten year agreement to produce and use RDF. Final approval was given by the City Council even though it was told landfill was the cheaper alternative. The decision was influenced by the fact that a new landfill would be required, which would encounter a great deal of public resistance, as well as by the fact that energy recovery was more in line with the environmentally conscious tenor of the times.

The RDF processing system is only operating at 50 percent capacity because of the lack of markets for the fuel (see Figure 10–2). This lack of market for the RDF capacity of the plant is one reason that the cost per ton of the system is so high. To remedy this problem, the City has been working with Oscar Mayer & Company on the use of RDF as a supplemental fuel with coal. After extensive testing, burning of RDF consistently at Oscar Meyer came on-line in June 1983. Other customers, together using approximately 4 percent of the RDF produced in 1983, include the Heil Co., Northern States Power Co., Institute of Gas Technology, and Boeing Engineering and Construction.

Economics

Capital costs incurred by the City of Madison amounted to more than $3.7 million (see Table 10–6). This can be divided into two groups: $2.9 million for the processing plant and $1.0 million for the receiving facility. MG&E also invested $0.5 million in modifications to its boilers, but Madison is reimbursing MG&E for this expenditure on a five year payback schedule. The project was financed through ten year bonds at 5.5 percent interest along with a small Community Development Block Grant.

Table 10–7 presents a summary of the system's annual economics for 1984. Costs for the RDF processing system and for MG&E and Oscar Mayer receiving stations are shown separately. As noted earlier, the magnetic separator and the trommel are privately owned by Madison Magnetic Operations (MMO) and Madison Solid Waste Fuel Company (MSWF), respectively. The City bears the cost of operating these systems but then recovers a portion of the costs through charges to the owners. The City expects to receive a percentage of the revenues obtained from the sale of ferrous metal. However, MMO is only obtaining limited revenues from metal recovery, and this is reflected in the revenue split going to the City. In addition, the City pays an annual fee to MSWF based on a sliding-scale throughput for the use of the trommel.

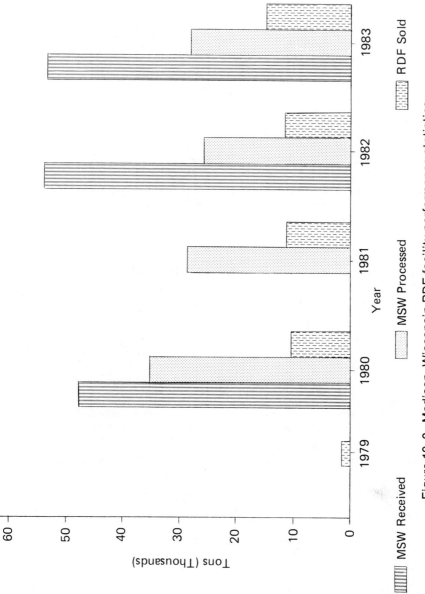

Figure 10-2. Madison, Wisconsin RDF facility performance statistics.

Table 10–6. Madison, Wisconsin RDF Facility Capital Costs.[a]

CITY OF MADISON

Processing Plant		
Direct Construction Costs		
Building and Grounds	1,024,976	
Process System[b]	1,259,183	
Indirect Costs		
Start-up, Engineering	438,606	
Subtotal		2,722,765
Receiving Plant		
Direct Construction Costs		
Building and Grounds	282,065	
RDF Feed System	544,471	
Indirect Costs		
Start-up, Engineering	228,135	
Subtotal		1,054,671
Total		3,777,436

MADISON GAS AND ELECTRIC

Receiving Plant Boiler Modifications	500,000	
Total		500,000
Grand Total		4,277,436

[a]Cost data are in 1978 dollars.
[b]Excludes cost of ferrous metal recovery system and trommel device. Both units are privately owned.

The revenues shown include the portion of the operating costs for the ferrous separator and trommel that are charged to MMO and MSWF. The net cost of the full system is $23.18 per input ton. Madison feels that promising economic projections based on future energy revenues, together with the environmental attractiveness of this approach, justify the system's operation.

1984 Project Summary[2]

The City of Madison and Madison Gas & Electric Company's Energy Recovery Program completed its fifth year of operation on December 31, 1984. During the Energy Recovery Program's first five years of operation, Madison Gas & Electric Company co-fired 58,325.99 tons of prepared

2. This section is excerpted from: *Energy Recovery Program Annual Report 1984,* City of Madison, Department of Public Works, April 10, 1984.

Table 10–7. Madison, Wisconsin RDF Facility Financial Summary 1984.

OPERATING EXPENSES	TOTAL	PER TON[a]	
RDF Processing System			
Labor (not including hauling)	364,750	6.11	
O&M	342,664	5.74	
MSWF Payment	189,922	3.18	
Hauling[b]	227,322	3.80	
Debt Service	196,237	3.29	
Disposal Fees[c]	142,245	2.38	
Subtotal	1,463,140	24.50	
MG&E Receiving Station			
O&M	174,870	2.93	
Debt Service	87,488	1.46	
Subtotal	262,358	4.39	
Oscar Mayer Receiving Station			
O&M	20,933	0.35	
Debt Service	66,101	1.11	
Subtotal	87,034	1.46	
Subtotal Operating Expenses	1,812,532	30.35	
OPERATING REVENUES	TOTAL	PER TON[a]	PER TON[d]
RDF Sales	350,791	5.87	19.24
Ferrous Metal Sales	543	0.01	0.57
MSWF Back Charges	7,162	.12	
MMO Back Charges	3,829	.06	
Tipping Fees/Other	66,001	1.11	
Subtotal Operation Revenues	428,326	7.17	
Net Cost of System	1,384,206	23.18	

[a] Based on 59,710 tons of raw solid waste input to the system in 1984 (includes 2,814 tons of large unprocessible items).

[b] Hauling includes labor and equipment costs for residue disposal, RDF transfer, and haul of unprocessible wastes to landfill.

[c] Disposal fees include 13,930 tons residue for system and 4,630 tons primary shred to landfill.

[d] Average revenue value, dollars per ton of product sold.

municipal solid waste (refuse-derived fuel) with coal at its Blount Street Generation Station. The refuse-derived fuel was fired in Madison Gas & Electric Company's boilers No. 8 and No. 9. Boilers No. 8 and No. 9 are Babcock and Wilcox 50 M pulverized coal boilers modified to co-fire RDF and coal for electrical power generation.

In 1982, the City of Madison and Oscar Mayer & Company, a national meat processor whose corporate offices and largest meat packing plant are located in Madison, completed negotiations on a long-term agreement to co-fire RDF for steam production at the firm's Madison plant. The agreement evolved from a three-year EPA demonstration project to test fire RDF in an industrial boiler.

During the early portion of 1983, modifications to the existing receiving and feed facility located at Oscar Mayer & Company were completed. Shakedown of the new and larger equipment began in mid-June 1983. During 1984, an additional auger drive motor was added to compliment the feeding mechanism. Through 1984, 3,983.26 tons of refuse-derived fuel were co-fired at this location in boiler No. 5 (June 1983 marked the beginning of the negotiated agreement between the City of Madison and Oscar Mayer & Company and the start of revenue to the City from the sale of RDF). Boiler No. 5, at Oscar Mayer & Company, is a Wicks Traveling Grate Spreader Stoker, with a capacity of 125,000 pounds of steam per hour. Steam generated is used for plant processing and also electrical power generation.

The Energy Recovery Program's fifth year of operation continued to add valuable operating experience at both the City's Resource Recovery Plant and Madison Gas & Electric Company's RDF and Feed Facility. The increased operating experience resulted in the increased availability of most of the energy recovery system components in 1984. The actual tons of RDF, co-fired with coal, at Madison Gas & Electric Company remained constant in 1984. In addition, the added experience aided appreciably in increasing operating efficiency at the Oscar Mayer & Company Receiving and Feed Facility.

Project Evaluation

Two important lessons can be gained from Madison's experience with energy recovery. The first lesson is specific to the RDF approach to recovery. The solid waste processing system used at Madison has proven to be a major improvement in the production of RDF. Elimination of heavy shredding, as a first step in processing, significantly reduced the embedding of inorganic material in the combustible fraction, resulting in a decline in the ash content of the RDF. In previous RDF systems, a high ash content has been a serious obstacle to developing markets. The experience to date in Madison demonstrates that RDF can provide a significant, reliable energy source for electrical utilities.

Secondly, Madison's experience illustrates that a system does not have to provide the cheapest available disposal alternative in order to be implemented. Madison's long previous experience with shredding led it to RDF as an energy recovery option. This fact, coupled with the market's need for a fuel compatible with coal and the public's resistance to landfilling, contributed to making the system a reality.

PITTSFIELD, MASSACHUSETTS

Introduction

As with many medium-sized communities, the early 1970s found the City of Pittsfield, Massachusetts confronting a dilemma: the need for additional landfill space in the face of stronger landfill environmental regulations. The existing landfill was encroaching on the City's only industrial park, and an extensive search for a new site found no politically acceptable location. After five years of study, the City selected Vicon Recovery Associates to design, build, and operate a 240 TPD waste-to-energy facility to serve the City and surrounding communities. This facility, located on the eastern edge of Pitts-field, has been in commercial operation since mid-1981, selling steam to a local paper manufacturer (Crane and Company).

The City of Pittsfield, with a population of approximately 52,000, is located in central Berkshire County in western Massachusetts. Employing over 8,000 people, General Electric is the major employer in the area, with plastics and pulp and paper manufacturers being other key employers. The City serves as the manufacturing and retail center for the area and is located approximately three hours' driving time from New York City and Boston.

Background

In 1973, Pittsfield's landfill, located adjacent to the Housatonic River, was nearing capacity. Future expansion was limited because the landfill was situated on the southern edge of the City's industrial park, with further ex-pansion restricted by railroad tracks to the south and the river to the east. In addition, poor operation of the landfill had resulted in an overriding negative impression by the surrounding industrial and residential neighbors. In February 1974, the Mayor responded to these concerns by appointing a Solid Waste Commission responsible for evaluating the available disposal alternatives. The first major action of the Commission was to negotiate for additional landfill space within the industrial park. This was accomplished with the agreement of all that no further extensions would be requested and an alternative site or disposal method would be found within six years.

Initial evaluation efforts by the Commission focused on recycling methods and a search for new landfill sites. It was clear after a year of exten-sive analysis and a survey of the interest of Pittsfield's residents in a source separation program that no politically acceptable landfill site could be found and recycling efforts would not sufficiently reduce the amount of solid waste generated. At the Commission's recommendation, future study efforts focused on evaluating waste-to-energy technologies, and Commission membership (originally composed of several environmentalists) was reshaped in 1975 to include several technically oriented members. After considering various technology options, the Commission identified refuse-derived fuel (RDF) and mass firing of refuse to produce steam as the two technologies warranting further investigation. Simultaneously, a search was made for prepared fuel and steam markets. This led to the eventual elimination of RDF, as no local industry indicated an interest in modifying its boiler system to accept RDF. Two potential markets for steam were identified in the preliminary study issued at the end of 1975. Commission recommendations in this study were:

- A refuse-to-energy plant would be a feasible disposal alternative.
- The facility should be privately owned and operated.
- The City should continue to pay to dispose of refuse (it was assumed that the waste supply could be maintained if there was no cheaper disposal alternative for the haulers).
- The project should have a reasonable tax impact.

Three separate estimates of the cost to operate a regulated landfill were prepared with per-ton tipping fees of $11.55 to $14.00. As this would be a cost the City would have to bear if an energy recovery facility was built, $12.00 to $14.00 per ton was felt to be an acceptable disposal fee at the proposed facility. It was recognized that with the increasing value of energy, the net disposal fee should not increase as fast as the cost of landfill. The project was accepted politically on the basis that, over the financial life of the project, the total cost of disposal would be less than landfill.

These recommendations were verified by an engineering consultant engaged to evaluate project feasibility. In early 1977, the consultant's report, confirming the economic feasibility of a modular controlled air system, also cautioned the City that this type of technology had less history of success than the many waterwall units found in Europe.

Markets

During 1977, the City of Pittsfield formulated an implementation strategy. A regional survey had been conducted to determine whether surrounding communities were interested in joining the project. Since few of the neighboring communities were facing an immediate landfill crisis, sufficient interest could not be generated to expedite project implementation. The Commission also felt that its solid waste problem was so immediate that there was no time available to overcome the political difficulties involved in obtaining a multijurisdictional agreement.

However, the regional survey had indicated that if one allowed for the lead time necessary to develop a waste-to-energy project, several surrounding communities would be facing landfill closures by the time a City sponsored facility was in operation. The City elected to proceed with the development of a facility which would (1) have sufficient capacity to accommodate the waste from surrounding communities and (2) be sized to maximize the ability to sell steam.

A weighing program was instituted which continued until the spring of 1979. This allowed refinement of the tonnage figures through contract negotiations. Because a 1950 vintage municipal incinerator sat idle within sight of the landfill, the Commission also spent a considerable amount of time overcoming political opposition to what was viewed as an unproven technology. The City had built an incinerator in the 1950s—at a cost of one million dollars—which had been in service for only a few years when a combination of maintenance, operational, and environmental problems forced its closing. It served as local evidence to support the unfavorable reports on U.S. experience with municipal operation of incinerators. The presence of this abandoned facility also influenced the City's decisions on project structure. The City's decision to issue a Request for Proposals (RFP) was conditioned upon attracting a private contractor to operate the facility. This position was supported by Crane and Company's reluctance to deal with a municipality. The project chronology is summarized in Table 10-8.

In March 1978, the City issued the RFP, soliciting proposals from private industry to design, construct, and operate a resource recovery facility. The RFP stated the City's willingness to enter into a put-or-pay contract for delivery of solid waste, to provide a site for the facility as well as a residue and backup landfill site, and to act as the vehicle for tax-exempt financing of the project. The RFP also identified Crane and Company as the steam customer and provided waste quantity data.

Three firms responded to the solicitation. After initial evaluation, two of these proposals were determined to be complete and responsive. Following a two month period of technical and financial evaluation, the City selected Vicon Construction Company, Inc. of Lincoln Park, New Jersey to enter into negotiations. Although this would be Vicon's first waste-to-energy facility, it was selected on the basis of stronger financial assets and potentially lower disposal service cost to the City.

The negotiated procurement process called for in the RFP stipulated that negotiations with the selected firm would continue until they were no longer considered to be "in good faith." If that event had occurred, negotiations would have commenced with the alternative bidder, Consumat Systems, Inc. of Richmond, Virginia.

During negotiations with the City, Vicon was responsible for negotiating a steam purchase agreement with the steam customer, Crane and Company.

Table 10-8. Pittsfield, Massachusetts Mass Burn Facility Chronology.

1974	Solid Waste Commission formed
	6.1 acres of additional landfill secured
1975	Evaluated technologies, targeted resource recovery
	Request for $40,000 (consultant) to prepare feasibility study of energy and material recovery from solid waste
1976	Landfill weighing program
	Market analysis: General Electric Co.
	Crane & Company
	Feasibility of material and energy recovery from solid waste study
	Survey of regional interest
	Targeted starved air system
	Continued search for new landfill
1977	Crane signs letter of intent
	Commission recommends proceeding with energy recovery
1978	Landfill plan prepared
	RFP issued in March (full service)
	Pittsfield Industrial Development Finance Authority ownership to take advantage of tax-exempt bonds
	Vicon selected and enters negotiations with City
1979	Initial contract signed with Vicon
	Vicon negotiates steam contract with Crane
	City/Vicon contract amended for design and bond issue
	Sept. 6—$6.2 million bond issue sold
	Construction started
	Resource Recovery Commission created by City ordinance
1980	Construction
1981	Feb. 6—first refuse burned
	June 8–June 13—preliminary commitment test passed

The basic agreements were signed in February 1979, with amendments made until the date of bond issuance in August 1979. Additional efforts during this period included the sale by Crane to the City of approximately 5 acres for the facility site, and the formal establishment of the Pittsfield Industrial Development Finance Authority to be the legal owner of the facility. (Vicon was the owner of the tax benefits under a lease arrangement with the Authority.)

The prevailing philosophy through negotiations was one of a cooperative venture in which all participants would get a "fair deal." Key contract elements contained in the Solid Waste Disposal and Resource Recovery Agreement negotiated between the City and Vicon included:

- Vicon would dispose of all waste which historically went to the landfill for a period of 15 years. Further, Vicon must operate and maintain the facility in such a manner that it can receive and process up to 240 TPD.
- The City would guarantee delivery of 44,000 TPY (tons per year).
- The tipping fee for waste delivered by the City would be determined by formula, with the City assuring a minimum.
- Vicon was obligated to complete construction by September 15, 1980 unless the time was extended due to an act beyond the control of either party.
- Profits from the plant would be split 50/50 between the City and Vicon.
- The facility would be required to meet a preliminary commitment test before the City's obligation to deliver and pay for waste disposal and Vicon's obligation to accept and process the waste would become absolute.
- Final acceptance of the facility by the City would occur when the project demonstrated the ability to (1) incinerate waste at its guaranteed plant capacity of 240 TPD, (2) generate salable steam, and (3) meet all other specifications.
- After the first year of operation, Vicon could enter into waste supply contracts with private haulers or other municipalities as long as these did not impair the facility's ability to accept City waste.
- The City would own the plant and property and, in turn, lease them to Vicon through the Industrial Development Financing Authority.

The 15 year steam purchase agreement negotiations by Vicon with Crane included the following provisions:

- *Construction:* Vicon would construct all steam-producing facilities as well as line and appurtenances necessary for steam delivery and would provide steam-metering stations. Auxiliary boilers and standby electrical power equipment would be installed to ensure steam delivery.
- *Steam delivery:* Vicon would be required to sell and deliver, and Crane to accept and purchase, at least 700,000 pounds of steam per day at a rate of at least 20,000 pounds per hour for 240 Crane work days. Greater quantities of steam may also be sold if Crane is able to use them (the project is expected to provide about 60 percent of Crane's total steam requirements).
- *Payments:* The price of the steam would be based on Crane's costs for No. 6 fuel oil discounted by a negotiated rate which will increase from the first year of sales (15 percent) to the fifteenth year (25 percent), adjusted according to the boiler efficiency of Crane's existing steam-generating equipment.

- *Termination:* The agreement could be terminated by Crane if the facility was unable to furnish steam by December 1, 1981.

After commencement of the agreement (with notice to Vicon), Crane may terminate if Vicon fails to deliver the required steam, breaches the agreement, or goes bankrupt. Also, either Crane or the company may terminate if there is a material change in law or if Vicon's contract with the City is terminated before steam is being supplied. Vicon, upon notice to Crane, may terminate the agreement if Crane fails to pay for steam, breaches the agreement, or goes bankrupt. Either party may terminate without liability if either the project or Crane is affected by an event beyond its control.

As the primary manufacturer in Pittsfield, General Electric (GE) was given first consideration as an energy market. Preliminary analyses indicated that the quantity of steam to be produced by a waste-to-energy facility would meet only 10 percent of the GE winter load. Although the summer steam demand was 60,000 pounds per hour, GE's initial projections indicated that Pittsfield's waste could be converted to approximately 40,000 pounds of steam. As GE's smallest boiler is rated at 100,000 pounds per hour, installation of a new, smaller, backup boiler would be required. Thus, economic utilization of the steam was not viewed as feasible by GE officials. In addition, numerous institutional issues inherent in working with a large company were viewed as negative considerations and led to the elimination of GE as a potential energy market.

Crane and Company of Dalton, Massachusetts (a manufacturer of quality papers, including all currency for the U.S. government) was identified as the preferred energy market. Crane's average weekday steam load varies from approximately 90,000 pounds per hour in the winter to 60,000 pounds per hour in the summer. Crane's four No. 6 oil-fired boilers gave considerable flexibility for matching steam delivery capabilities of a waste-to-energy facility.

Technology

The project site, a 5 acre parcel located on Hubbard Avenue, is on the eastern edge of Pittsfield, across the Housatonic River from one of five Crane mills. Facility layout provides for traffic flow separating commercial traffic from private vehicles, with private resident disposal bins recessed in front of the building. All other vehicles delivering waste pass over a scale adjacent to a free-standing office and scale building.

The processing building is approximately 30,000 square feet in area and 31 feet high, with six doors to receive incoming waste deliveries. The plant, constructed of modular precast concrete panels, includes a refuse storage pit with a 500 ton capacity and a 5 ton crane to transfer stored waste from the pit to the tipping floor. Vehicles may also dump waste directly onto the tipping floor. Front-end loaders then transfer the waste into the incinerator receiving hopper.

Vicon holds an exclusive license to market and manufacture modular controlled air MSW incinerators designed by Enercon Systems, Inc. of Cleveland, Ohio. Enercon Systems, Inc. is a high-technology system engineering company specializing in solid waste incineration, energy conservation, resource recovery, and air pollution control. In addition to providing system engineering for this project, Enercon manufactured the control system and several other components.

Three 120 TPD furnaces were installed, with two furnaces to be on-line and one standby. The primary furnace chamber is a rectangular, stepped hearth design with precast refractory-lined roof and side panels. The preassembled panels are positioned on-site, with the floor of each unit lined with refractory brick. Waste is hydraulically conveyed from the $4 \times 6 \times 8$ cubic foot loading hopper onto the first hearth of the primary chamber. Features of the units include controlled overfire and underfire airflows and water cooling of steel components. Waste and ash material is hydraulically conveyed through the primary chamber by water-cooled rams and deposited in a water-filled quench trough common to all three units.

Combustion within the primary chamber is regulated by the introduction of air, with the system maintained in an excess air environment. Gases leaving the primary chamber enter the secondary chamber where combustion is completed. Burners fired by either natural gas or oil at the end of the secondary chamber can be used to maintain constant gas temperature, although use of these burners has been negligible to date. Hot gases passing from the secondary chambers are received in a shared tertiary chamber or manifold which then directs the gases to the waste heat boilers.

Two waste heat boilers manufactured by Bigelow are installed side by side and can be operated dually or individually. The normal flow is from any one furnace to either waste heat boiler or from any two furnaces to both boilers. When energy recovery is not possible, heat is automatically dumped through a bypass stack; however, gases exiting during this dump stack mode are considered to be in violation of state air pollution standards. Gas flow within the system is controlled by large dampers capable of isolating the various system components.

The waste heat boilers, rated at 35,000 pounds per hour each, are designed to operate with flue gas temperatures up to 1,700°F and generate superheated steam at 250 psi. During normal operations, approximately 30,000 pounds of steam per hour is delivered to Crane.

Exiting gases pass through one of two "electroscrubbers." This is a relatively new type of air pollution control equipment, designed by Combustion Power Equipment Co., a subsidiary of Weyerhaeuser. Flue gases pass through a circular bed of $\frac{1}{4}$ inch gravel containing an electrostatic grid to enhance the capture of particles. The gravel is pneumatically conveyed to the top of the scrubbers, cleaned, and then returned to the top of the gravel bed, where it is exposed again to the flue gases. Cleaned gases leave through the middle stack of the circular gravel bed and up the stack.

The City/Vicon contract called for the City to provide a residue and bypass landfill to be operated by Vicon. After an unsuccessful search for a new landfill site, the City approached the Massachusetts Department of Environmental Quality Engineering (DEQE) with a proposal to cap the existing municipal landfill with two layers of ash and bypass material. A plan outlining this procedure and the transfer of operations was submitted to, and approved by, DEQE. In early 1980, the City initiated efforts to close one portion of the landfill while retaining a 6 acre parcel for ongoing operations until the facility completed the initial period of operation and passed all acceptance testing.

In February 1981, Vicon began accepting waste for testing, and on April 13, 1981, the City officially closed its landfill operation. Since this date, Vicon has accepted all City waste.

A four month initial period of operation was defined by contract. During this period, Vicon was required to pass a preliminary commitment test demonstrating the plant's ability to legally incinerate at least 170 tons of acceptable waste per day for five consecutive days. The facility passed this test June 8, 1981–June 13, 1981, securing the City's obligation to deliver waste and pay a service fee, and obligating Vicon to accept and process the waste. A technical performance test demonstrating the facility's capability to process 240 TPD and generate steam was conducted and passed in October 1981.

During the first full year of operation (April 1981–March 1982) just under 42,000 tons of solid waste were processed. Initial steam delivery problems resulted in reduced steam sales. These problems were caused by refuse-derived steam being introduced to Crane's steam distribution system at a point downstream from the power plant. However, steam line modifications have since remedied this situation. As would be expected with the first application of a technology, the operator initiated a series of system modifications based on experience gained during the initial period of operation. Modifications have included reconstruction of the refractory line for the primary combustion chambers, improved airflow control, and redesign of the pneumatic components for the air pollution control equipment.

Minor problems were also experienced with the hydraulic controls for the incinerator rams which required different pressures for different functions.

Modifications in the sychronization of furnace operations and pressure changes have corrected the situation.

The refractory problems caused by a design flaw in securing the refractory liner to the metal shell of the incinerator have been corrected at the expense of Vicon by providing an air insulation layer. Slagging problems developed within the first hearth of the primary chamber because of cool air entering the chamber around the loading ram. Again, design modifications initiated by Vicon have corrected this problem.

A sludge press was installed in the plant in 1983. Dewatered paper sludge from Crane is conveyed into the storage pit for mixing with MSW and introduction to the incinerators.

Economics

Influenced by the reluctance of Pittsfield's politicians to finance another incinerator and Crane's reluctance to participate in a City operated steam plant, the Solid Waste Commission recommended that the energy recovery facility be privately financed and operated. Through the RFP, the City indicated its endorsement of this position and its willingness to act as a vehicle for project financing, with the selected vendor required to have sufficient financial capabilities to secure all necessary financing. By structuring the project with the Pittsfield Industrial Development Financing Authority (IDFA) holding title to the property, tax-exempt revenue bonds guaranteed by Vicon could be issued.

With the intention that project revenues would be sufficient to cover debt service and operating costs, the five project contracts were structured to serve as security features for project financing with (1) the operating contract between the City and Vicon guaranteeing payment of a service fee for 15 years, (2) the lease agreement between the IDFA and Vicon authorizing Vicon to act as owner, (3) the steam purchase agreement between Crane and Vicon stipulating a minimum steam purchase price for 15 years, (4) the trust indenture securing the bonds by mortgaging the facility, and (5) the bond provided by Vicon to the bondholders guaranteeing bond payment if project funds are insufficient.

The City holds legal title to the project so long as the bonds are outstanding. Vicon, however, derives the benefits of constructive ownership, including certain tax credits and deductions. The original bond proceeds were insufficient to complete the construction, and the company paid the costs of completion (the company initially expected to make an equity contribution to the project of only about $500,000). The company's obligation to make rental payments equal to debt service on the bonds is absolute and unconditional.

All payments under the service contract and the steam purchase agreement, as well as the company's rental payments, are made directly to the trustee. The trust indenture creates four trust funds with the trustee: the bond fund, the debt service reserve fund, the construction fund, and the revenue fund. All disposal fees payable by the City to Vicon Recovery Associates and amounts due from Crane under the steam purchase agreement are deposited in the revenue fund for application to the other funds as specified. Funds remaining after debt service payments are passed to Vicon for the company's payment of approved operating and maintenance costs. Any additional revenues are distributed between the City and Vicon based on a profit sharing formula which allocates approximately the first $100,000 to Vicon as a management fee, with remaining funds to be split fifty-fifty with the City. Funds owed the City are applied against the next year's service fee.

Project cost estimates provided by Vicon and verified as feasible by a third party review for bond issuance were used to set the bond size. In addition, bond costs and a debt service reserve fund were calculated into a total bond issue of $6.2 million, of which $4.7 million was to be available for construction. Vicon initially contributed an additional $500,000 in equity. When a number of adverse factors including bad publicity relative to resource recovery in general, a sluggish municipal bond market, and the high leverage characterizing the transaction threatened the financeability of the project, the City lent its financial credibility by providing a secondary guarantee to the bondholders in the event Vicon became insolvent. Thus in September 1979, the $6.2 million bond issue was closed with interest rates of different maturities ranging from 6 to 7 1/2 percent. A breakdown of the bond issue funds is provided in the first column of Table 10-9.

The City's base service fee could then be calculated from the bond amortization schedule. In 1981 this service fee was $11.68 per ton. Subsequent adjustments were made to account for property tax payments and profit sharing credits. During periods of contract default by Vicon, the City has the unconditional obligation to pay a service fee sufficient to cover debt service. Based on information provided by Vicon, project capital costs have totaled $10.9 million, with only $4.7 million available from the bond issue. Table 10-9 presents the actual project costs for comparison with the original projection. As set forth by contract, Vicon provided for the additional financing, although negotiations were held with the City on the basis that a portion of the additional cost can be attributed to project improvements that will increase energy recovery. Vicon has acknowledged that $3.6 million of its additional equity was provided by a private financing placement.

Table 10-9. Pittsfield, Massachusetts Mass Burn Facility Capital Costs.

	PROJECTED (1979)[a] ($1,000)	ACTUAL ($1,000)
A. *Equipment and Building*		
Land/Site Development[b]	265	1,063
General Equipment Construction/Erection	867	902
Loader(s)/Boiler(s)	210	590
Incinerator(s)/Boilers(s)	2,300	6,650
Ash Removal System	35	218
Air Pollution Control	335	862
Stack(s)	—	—
Cogeneration Equipment	—	—
Pipeline	—	64
Misc. Equipment and Controls	301	505
Subtotal Equipment and Building	4,313	10,854
Vicon Fees and Services	892[c]	—[d]
SUBTOTAL	5,205	10,854
B. *Fees and Services*		
Debt Service Reserve Fund (less interest earnings)	425	425
Capitalized Interest	653	653
Bond Issuance Costs	207	207 (est.)
Legal	40	40 (est.)
Financial Consultant	170	170 (est.)
Authority Development	—	—
SUBTOTAL	1,495	1,495
TOTAL	6,700	12,349

[a] Based on August 1979 bond prospectus.
[b] Land costs amounted to $1.00.
[c] Includes design/construction management, start-up costs, contingencies, and permits/approvals.
[d] These cost items are included in the equipment and building costs.

First year steam sales revenues were less than projected due to initial delivery problems. Additionally, the facility was undergoing start-up through June 1981. The 1982 and 1984 operational records are presented in Table 10-10. In 1982, a total of 61,968 tons were processed, generating $771,330 in tipping fee revenues and $1,092,540 in steam sale revenues. In 1984 the project sold an average of 29,500 pounds of steam per hour (8,400 hours), thereby meeting project goals. Figure 10-3 illustrates 1983 performance.

Initial project economics were based on the facility processing only the City's waste stream. However, contract provisions allowed Vicon to con-

Table 10–10. Pittsfield, Massachusetts Mass Burn Facility Project Economics (1982 and 1984).[a]

	1982 ($1,000)	1984 ($1,000)
I. *Revenues*		
Steam	1,093	1,467
Tipping Fees	771	904
Other	128	—
Subtotal	1,992 (32.15/ton)	2,371 (33.58/ton)
II. *Expenses*		
Labor	513	605
Utilities	187	179
Refuse/Residue Landfill[b]	N/A	N/A
Spare Parts/Supplies/Maintenance	301	379
Administration/General	102	133
Subtotal	1,103	1,296
Debt Service and Interest Expense	666	666
Subtotal	1,769 (28.50/ton)	1,962 (27.80/ton)
III. *Net Revenues*		
Annual	223 (3.60/ton)	409 (5.80/ton)

[a] A total of 61,968 tons of waste was processed in 1982 and 70,612 tons in 1984.
[b] Included in maintenance budget.

tract with surrounding communities for additional waste after the first year of operation. Several contracts have been signed with surrounding towns, and contracts have been secured with area private haulers for an additional interruptible supply of waste.

Project Evaluation

Vicon has taken the position that the Pittsfield facility has operated with a 100 percent availability since March 28, 1981. This is based on a 50 week per year schedule with a two week planned outage in July. To a large degree this is possible because of the availability of three furnaces, only two of which are required to be on-line at any one time. This system redundancy was required by the City to offset the risk of selecting a technology that was new at the time and a vendor with a limited track record.

The fact that the facility capital costs exceeded initial projections by more than 100 percent can be attributed to numerous factors, not the least of which was the impact of an inflationary economy on the construction in-dustry. The significance to future project structuring is that contractual pro-

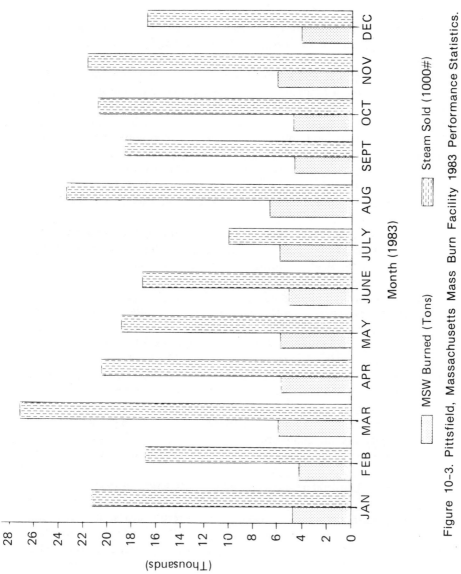

Figure 10-3. Pittsfield, Massachusetts Mass Burn Facility 1983 Performance Statistics.

visions protected the City from responsibility for the cost overruns unless the contractor, Vicon, could identify project benefits resulting from an increase in energy recovery.

Summary of 1984 Operations.[3]   *Refuse Supply.*   Pittsfield's waste supply was up 4.2 percent from 63,135 in 1983 to 65,804 tons (1984). This is believed to be a weather-related phenomenon. Along with the expected increase due to "garage and attic" cleanup in the spring, it is theorized that the late May flood brought in "cellar remains" as well during June and July; the weekly average during these months showed a 6 percent increase over the year before. More significant was the mild weather throughout the fall, especially on the weekends. The fourth quarter was up 12 percent from 1983; the weekly average for October was up 16 percent over 1983.

The City contractor, Clean City, brought in 1,765 tons more than in 1983, an increase of 12 percent. This represents more than one-half of the increased tonnage, significant in view of the fact that it hauls only 30 percent of Pittsfield's waste.

Non-Pittsfield waste was about the same, accounting for 20 percent of the total. Crane sludge (4,808 tons) was 8 percent lower than the year before, but the overall total for 1984 was 70,612 tons of waste and sludge delivered—the highest total since commencement of operations.

The total quantity of bypassed waste sent to the landfill was 7 percent less than the 1983 quantity, even though it was significantly higher during the plant's scheduled shutdown. It is noted that the refuse sent to landfill (other than during the two week plant shutdown) was just over 1 percent of the total waste (less than one-half of the previous year). The total raw waste sent to landfill was 3,493 tons, or 5 percent of the waste received. Of this raw waste, about 4 percent (2,659 tons) occurred during plant shutdown, i.e., two weeks in July when Crane is also down for vacations.

Overall, the quantity of waste incinerated was up about 4 percent from the previous year; it is anticipated by the operator that this might be increased another 5 percent as continued fine tuning and modifications take place.

Steam Sales.   The total quantity of steam sold to Crane in 1984 needs interpretation because the auxiliary fossil fuel boiler at the facility was used more extensively during 1984 to increase the sale of refuse steam. Whereas total steam sold to Crane increased from 232.6 million pounds (1983) to 248.4 million pounds (1984), refuse-derived steam actually increased from 231.3

---

3. This section adapted from *Summary of Operations (Calendar Year 1984)*, Vicon Recovery Associates.

million to 241.9 million pounds, an increase of 4.6 percent. This is consistent with the increase in refuse burned.

At the end of 1984, plans were underway to move the auxiliary boiler to Crane and Company, which should allow better opportunity to sell more refuse-derived steam.

Plant Operations. Plant performance has been measured by several coefficients: availability, production, and efficiency. The standards established by Vicon Recovery Associates are 95 percent for availability, 5 million pounds of steam sold per week, and 2.2 pounds of steam made per pounds of waste burned. The average for the 1984 calendar year indicated that performance coefficients (based on 1.0 as meeting the goal) were 1.01, 0.95, and 0.99, for availability, production, and efficiency, respectively. The efficiency (i.e., pounds of steam sold per pound of waste burned) showed an average increase of 8 percent for the second half of 1984 over the first half and previous periods. The operator claims this was due to improvements made at plant shutdown and more equipment fine tuning. Production and efficiency levels were significantly down in December 1984 because of two factors: (1) the facility ran out of solid waste; (2) Crane (the energy market) was shut down from December 25 through New Year's Day.

During plant shutdown in 1984, the most significant maintenance accomplishment was rework of the recirculated flue gas system. Modifications to the underfire air system continue to be made, and all three furnaces now have the "final generation" cooling plates at the entrance hearth. The last part of the design of the underfire air system is to correct the potential corrosion problem when using recirculated flue gas.

A new water-cooled gasket retainer was installed on the explosion doors of two of the three furnaces. The operator feels this is a good solution to a problem which has plagued them since start-up.

In general, the Pittsfield energy recovery project can be considered successful and unique for several reasons:

- It represents the first application of industrial revenue bond financing for a small-scale resource recovery facility.
- The system to be utilized had never been demonstrated on a commercial basis.
- The company responsible for the design, construction, and operation of the project had never built or operated an energy recovery facility.

The significance of this project is further enhanced by a review of the chronology showing a five year development schedule once energy recovery was identified as a feasible solid waste management option. Furthermore,

Table 10–11. Pittsfield, Massachusetts Mass Burn Facility
Case Study Summary.

A. *General Characteristics:*

1. Location	Pittsfield, Massachusetts
2. Owner	Pittsfield Industrial Development Finance Authority
	City Hall
3. Procurement Approach	Full service
4. Design Capacity	240 TPD
5. Process	Modular, excess air (3 units @ 120 TPD, two on line at any one time)
6. Major Equipment Vendors	Furnaces, Enercon; waste heat boilers, Bigelow; air pollution control equipment, Combustion Power Company
7. Operator	Vicon Recovery Associates
8. Start-up Date	April 1981
9. Current Status	Full-scale commercial operation
10. Energy Market	Crane and Company (paper manufacturer)
11. Financing Mechanism	Tax-exempt Industrial Development Bonds plus private equity contribution

B. *Economics*

1. Capital Cost ($)

Industrial Revenue Bonds	6,200,000 (79)
Private Equity	6,149,000 (81)
Total	12,349,000
Escalated to 1981 dollars	13,575,000
Design Capacity (dollars/TPD)[a]	56,500

2. Operating Cost (1982)

	1982 DOLLARS	DOLLARS/TON
O&M	1,103	17.80
Debt Service and Interest Payments	767	12.40
Subtotal	1,870	30.20

3. Revenues (1982)[b]

Energy	1,093	17.60
Other	128	2.10
Tip Fees	771	12.40
Subtotal	1,992	32.10

4. "Actual" Net Revenue   121   1.90

C. *Performance*

1. Throughput vs Capacity (%): 240 TPD design—61,968 tons processed in 1982, 177 TPD throughput 74% capacity (50 weeks/year).

2. Technical Problems:
During Start-up: furnace refractory required reconstruction; redesign of air pollution control equipment pneumatics.
During Commercial Operation: steam delivery limitations; ash conveyor jamming; underfire air control; subsequent design modifications have been made.

3. Other Problems: Skyrocketing interest rates when bond issue hit market (City required to give limited guarantee); no politically viable new landfill site.

4. Future Changes: none planned; fine tuning.

[a] Based on design capacity of 240 TPD.
[b] Based on 41,894 tons processed.

the facility was constructed in an 18 month period in order to meet the City's deadline imposed by a very limited landfill life.

The technical problems encountered, while significant in themselves, point to the ability of a private operator to make the necessary modifications within the constraints of continuing commercial operation. A summary of this project is presented in Table 10–11.

BIBLIOGRAPHY

Chapter 1

Alter, H. and J. J. Dunn, Jr., *Solid Waste Conversion to Energy; Current European and U.S. Practices,* Marcel Dekker, Inc., 1980.

Bernheisel, J. F. *Resource Recovery Analysis Briefing Workbook,* prepared for the U.S. Department of Energy under Contract ES-76-C-01-3851, 1981 (co-author).

Bernheisel, J. F. "Nashville: A Successful Refuse-to-Energy Program," NCRR *Bulletin,* March 1979, Vol. 9, No. 1.

Brickner, R. H. "Case Studies: Four Small-Scale Refuse to Energy Plants." *21st Annual International Seminar.* GRCDA, Winnipeg, Manitoba, September 1983.

Brickner, R. H. (co-author). "Case Study Review: Small-Scale Waste-to-Energy District Heating Plants." *Small-Scale Combustion Systems: A Waste-to-Energy Workshop.* U.S. Department of Energy, Argonne National Laboratory, et al., Nashville, TN, July 1983.

Gershman, H. W. "The Southwest Facility," paper presented at the *Municipal Resource Recovery Conference,* Energy Bureau, Arlington, Virginia, October 1984.

Mantell, C. L. *Solid Wastes: Origin, Collection, Processing, and Disposal,* New York, John Wiley & Sons, Inc., 1975.

McEwen, L. B. "A Nationwide Survey—Waste Reduction and Resource Recovery Activities." U.S. Environmental Protection Agency, (SW-142)., Washington, D.C., 1977.

Meier, P. M., and T. H. McCoy. "Solid Waste as an Energy Source for the Northeast." Policy Analysis Division, National Center for Analysis of Energy Systems, Brookhaven National Laboratory, BNL 50559, Upton, New York, June 1976.

National Center for Resource Recovery. "Waste-to-Energy Compendium," AC01-76-CS-20167, U.S. Department of Energy, Washington, D.C., April 1981.

Paladino, A. E., et al. "Materials and Energy From Municipal Waste, V. 2. Working Papers: Resource Recovery and Recycling From Municipal Solid Waste." Office of Technology Assessment, Washington, D.C., 1978.

Peterson, C.W. (co-author). "Energy Recovery in Japan and Western Europe." *International Small-Scale Municipal Waste-to-Energy Conference.* Resource Recovery Report, Washington, D.C., February 1985.

Peterson C. W. (co-author). "Solid Waste Management: An International Perspective." *International Small-Scale Municipal Waste-to-Energy Conference.* Resource Recovery Report, Washington, D.C., February 1985.

Schoenhofer, R. F., M. A. Gagliardo and H. W. Gershman. "Fast Track Implementation of the Southwest Resource Recovery Facility, Baltimore, Maryland," In *Proceedings: 1982 National Waste Processing Conference,* New York, N.Y., May 1982.

Stearns, R. P. and J. Woodard. "The Impact of Resource Recovery on Urban Landfill Requirements." *Waste Age,* Volume 8, No. 1. January 1977.

Systech Corporation. "Systems Analysis for the Development of Small Resource Recovery Systems." U.S. Department of Energy, Washington, D.C., October 1980.

Wilson, D. G., ed. *Handbook of Solid Waste Management.* New York, Van Nostrand Reinhold Co., 1977.

Chapter 2

Gershman, H. W. "Solid Waste-to-Energy Overview and Status Report." *Proceedings of Twelfth Energy Technology Conference.* Government Institutes, Washington, D.C., March 1985.

Gershman, H. W. "Forces Behind Small-Scale Waste-to-Energy." *International Small-Scale Municipal Waste-to-Energy Conference.* Resource Recovery Report, Washington, D.C., February 1985.

Chapter 3

Aleshin, E. "Waste and Recycling Management." *Seminar on Materials Science and Processes.* Program for Technical Managers, Post College Professional Education, Carnegie-Mellon University, Pittsburgh, PA, May 1978.

Bernheisel, J. F. *Impact of Waste Variability on Energy Potential,* Bio-Energy '80—the Bio-Energy World Congress and Exposition, April 1980.

National Center for Resource Recovery. "Municipal Solid Waste . . . Its Volume, Composition and Value." *National Center for Resource Recovery Bulletin,* 3(2), 1973.

Peterson, C. W. and Gershman, H. W. *Compatibility of Newspaper Recycling and Energy Recovery.* Richmond, Virginia: Garden State Paper Company, December 1980.

Peterson, C. W. (co-author). "Resource Recovery and Waste Reduction." In *Third Report to Congress* (U.S. Environmental Protection Agency, Environmental Protection Publication SW-161). Washington, DC: U.S. Government Printing Office, 1975.

Peterson, C. W. "Reducing Waste Generation." In *Decision-Makers' Guide in Solid Waste Management* (U.S. Environmental Protection Agency, Publication SW-500). Washington, D.C.: U.S. Government Printing Office, 1976.

Skinner, J. H. "The Impact of Source Separation and Waste Reduction on the Economics of Resource Recovery Facilities." *Resource Recovery and Energy Review,* 5 p. March/April 1977.

Chapter 4

Bernheisel, J. F. "Business Arrangements for Energy Products from Solid Waste-to-Energy Projects." In *Proceedings of Twelfth Energy Technology Conference.* Government Institutes, Washington, D.C., March 1985.

Bratton, T. J. "Alternative Electricity Pricing Structures That Can Enhance the Development of Solid Waste Energy Recovery Projects." *Proceedings of Third Annual Resource Recovery Conference.* National Resource Recovery Association, Washington, D.C., March 1984.

Bratton, T. J. "Colleges and Universities as Markets for Energy from Solid Waste—An Overview of the Proposed Prince George's County/University of Maryland Project." *Association of University Architects Conference.* College Park, MD, June 1983.

Brickner, R. H. and L. Larochelle. "District Heating Applications of Small Sized Waste-to-Energy Plants," Gershman, Brickner & Bratton, Inc., Washington, D.C., July 1982.

Brickner, R. H. "Small-Scale Waste-to-Energy System Turbine Generator Systems Options." *International Small-Scale Municipal Waste-to-Energy Conference.* Resource Recovery Report, Washington, D.C., March 1985.

Brickner, R. H. "The Marketing of Solid Waste-to-Energy System Energy Products." *Market Development and Resource Recovery Conference.* U.S. Conference of Mayors, Washington, D.C., March 1983.

Garbe, Y. and S. J. Levy, "Resource Recovery Plant Implementation: Guides For Municipal Officials—Markets." U.S. Environmental Protection Agency, Washington, D.C., 1976.

Gershman, H. W. "How to Identify and Pin Down Markets for Outputs." *Solid Wastes Management,* May 1977.

Gershman, H. W. "Utility Negotiations." Paper presented at the U.S. Conference of Mayors, *Market Development and Resource Recovery Conference,* Washington, D.C., March 1983.

Gershman, H. W. (co-author). "Resource Recovery/District Heating Systems: Prospects and Problems." *70th Annual International Conference.* District Heating Association, Dixville Notch, NH, June 1979.

Gershman, H. W. "Marketing the Output." *Fifth National Congress on Waste Management Technology and Resource and Energy Recovery.* Dallas, TX, December 1976.

Gershman, Brickner & Bratton, Inc. "Cogeneration and Resource Recovery Opportunities." U.S. Environmental Protection Agency, Washington, D.C., February 1982.

Howard, S. E. *Market Locations for Recovered Materials.* U.S. Environmental Protection Agency, SW-518, U.S. Government Printing Office, Washington, D.C., 1976.

Liss, G. B. and L. Larochelle. "PURPA, Cogeneration and Resource Recovery." In *Proceedings: 1982 National Waste Processing Conference,* New York, N.Y., May 1982.

Chapter 5

GENERAL

Aldrich, R. and Rofe, R. "Small Resource Recovery Project Gets Disposal Revenue Bond Financing." *Waste Age,* January 1980.

Alvarez, R. "Status of Incineration and Generation of Energy from Thermal Processing of MSW." In *Proceedings: 1980 National Waste Processing Conference,* ASME, Washington, D.C., May 1980.

Brickner, R. H. "Evolution of the Small-Scale Incinerator Industry." *International Small-Scale Municipal Waste-to-Energy Conference.* Resource Recovery Report, Washington, D.C., February 1985.

Brickner, R. H. "Resource Recovery Vendors." *Resource Recovery Mayors Leadership Institute.* U.S. Conference of Mayors, Nashville, TN, October 1983.

Hecklinger, R. et al. "Oceanside Disposal Plant Improvement Program—Design, Construction and Operating Experience." In Proceedings: 1980 National Waste Processing Conference, Washington, D.C., May 1980.

Kirkpatrick, M. "Update on Nashville Thermal." In Proceedings: 1980 National Waste Processing Conference, Washington, D.C., May 1980.

Levy, S. J. and H. G. Rigo. "Resource Recovery Plant Implementation: Guides for Municipal Officials—Technologies." U.S. Environmental Protection Agency, Washington, D.C., 1976.

Peterson, C. W. "Combustion Technologies for Converting Urban Solid Waste to Energy in the United States." *Proceedings of Conference on the Use of Biomass for Energy by Industry.* Federacao as Industrias de Estado do Rio de Janeiro; Rio de Janeiro, Brazil, May 1982.

Peterson, C. W. (co-author). "Municipal Solid Waste for Energy: A Technology Review." *Proceedings of Eleventh Energy Technology Conference.* Government Institutes, Washington, D.C., March 1984.

Peterson, C. W. (co-author). "Small-Scale and Low Technology Resource Recovery." *Proceedings of the Second World Recycling Congress.* Manila, Phillipines, March 1979. Also, *Proceedings of the Fifth Annual Research Symposium on Municipal Solid Waste: Resource Recovery.* Environmental Protection Agency, Cincinnati, OH, March 1979.

Scaramelli, A. B. "Energy Market Key to Project Planning." *Solid Waste Management,* Volume 25, No. 4, April 1982.

"Source Separation, Small-Scale Pyrolysis Works Wonders for the Town of Plymouth." *Solid Waste Systems,* 6(3):14, 20–22, May-June 1977.

Taylor, H. et. al. "NASA/Hampton Refuse Fired Steam Plant: A Municipal/Federal Cooperative Effort." In Proceedings: 1980 National Waste Processing Conference, Washington, D.C., May 1980.

MODULAR INCINERATORS

Brickner, R. and Harrison, B. "Small May Be Better: Modular Incineration Offers An Alternative to Municipalities Without Large Volumes of Waste." *Waste Age,* March 1980.

Brickner, R. H. (co-author). "Smaller May Be Better: Modular Combustion Offers an Alternative to Municipalities Without Large Volumes of Waste." *Waste Age,* March 1980.

Frounfelker, R. "Small Modular Incinerator Systems with Heat Recovery: A Technical, Environmental, and Economic Evaluation." U.S. Environmental Protection Agency, Washington, D.C., 1979.

Gershman, H. W. "An Approach to Determining the Economic Feasibility of Refuse-Derived Fuel and Materials Recovery Processing," *Journal of Energy Resource Technology* (Transactions of the ASME), Vol. 102, No. 77, June 1980.

Larochelle, L. and H. Gershman. "Resource Recovery and Codisposal in Auburn, Maine." In Proceedings: 1980 National Waste Processing Conference, Washington, D.C., May 1980.

Niemann, K. "Salem's Modular Combustion System." *National Center For Resource Recovery Bulletin,* September 1980.

"North Little Rock: A Case Study of Economic Development and Resource Recovery." U.S. Conference of Mayors, Washington, D.C., November 1980.

Peterson, C. W. "Modular Incineration with Energy Recovery." *Appropriate Community Technology Conference.* Washington, D.C., May 1979.

Whitmore, P., C. Morrison, and L. Larochelle. "The Economics of Resource Recovery in Auburn, Maine." In Proceedings: 1982 National Waste Processing Conference, New York, N.Y., May 1982.

REFUSE DERIVED FUEL

Barlow K. "Burning Refuse-Derived Fuel in a 50 MW Utility Boiler." Presented at the EPRI Seminar—Municipal Waste as a Utility Fuel, Ft. Lauderdale, Florida, January 1980.

Barlow, K. et. al. "Design, Evaluation and Operating Experience of The City of Madison—Madison Gas & Electric Company Energy Recovery Project." In Proceedings: 1980 National Waste Processing Conference, Washington, D.C., May 1980.

Bernheisel, J. F. "Test and Evaluation at the New Orleans Resource Recovery Facility," *Proceedings* of the Seventh Annual Research Symposium on Land Disposal of Municipal Solid and Hazardous Waste and Resource Recovery, U.S. Environmental Protection Agency, March 16–18, 1981, Philadelphia, Pennsylvania.

Bernheisel, J. F. *New Orleans Resource Recovery Facility Shakedown Report,* National Center for Resource Recovery, 1981.

Bernheisel, J. F. "Mid-Shakedown Evaluation of a Demonstration Resource Recovery Facility," *Municipal Solid Waste: Resource Recovery,* Proceedings of the Fifth Annual Research Symposium, U.S. Environmental Protection Agency, August 1979.

Brickner, R. H. (co-author). "Refuse Shredding—Performance, Testing and Evaluation Data." *Proceedings of 1978 National Waste Processing Conference.* American Society of Mechanical Engineers, New York, May 1978.

Bernheisel, J. F. *New Orleans Resource Recovery Facility Implementation Study—Equipment, Economics, Environment.* National Center for Resource Recovery, 1977.

Campbell, J. and Renard, M. "Densification of Refuse Derived Fuels: Preparation, Properties, and Systems for Small Communities," U.S. Environmental Protection Agency, Cincinnati, Ohio, January 1981.

Gheresus, P. et al. "Resource Recovery from Municipal Waste: The Ames System Experience With Economics and Operation." In Proceedings: 1980 National Waste Processing Conference, Washington, D.C., May 1980.

Renard, M. L. and R. T. Fretz, "Feasibility of Burning Refuse Derived Fuels in Institutional Size Oil-Fired Boilers," ES-76-C-01-3851, U.S. Department of Energy, Washington, D.C., April 1979.

Vanroyan, G. L. and D. D. Huxtable. "The Laval d-RDF to District Heating Systems." In Proceedings: 1982 National Waste Processing Conference, New York, N.Y., May 1982.

Chapter 6

Gershman, H. W. "Impact of Tax Benefits on Project Economics," presented at the GRCDA *Resource Recovery Conference,* Dallas, Texas, August 1982.

Skinner, J. H. "The Impact of Source Separation and Waste Reduction on the Economics of Resource Recovery Facilities." *Resource Recovery and Energy Review,* 5 p. March/April 1977.

Sussman, D. *Resource Recovery Plant Implementation: Guides For Municipal Officials—Accounting Format.* Washington, D.C., U.S. Environmental Protection Agency, 1976.

Bratton, T. J. "RCRA Hazardous Waste Regulatory Impacts on Municipal Waste Incineration." Presented at the *Air Pollution Control Association Specialty Conference* on Waste Treatment and Disposal Aspects: Combustion and Air Pollution Control Processes, Charlotte, N.C., February 1981.

Berman, E. B. and E. A. Counihan. "An Economic Approach to Solid Waste Management Planning." Sponsored by the MITRE Corporation, MTP-158, Bedford, Massachusetts, March 1975.

Chapter 7

Peterson, C. W. "Cost Assessment of Biomass Conversion Technologies." *Proceedings of 9th Energy Technology Conference.* Government Institutes, Washington, D.C., February 1982.

Autio, A. E., et al. "Resource Recovery Implementation: An Overview of Issues." Metrek Division of the MITRE Corporation, M78-232, Bedford, Massachusetts, 1978.

Dawson, R. A. "Facility Financing Options." *Solid Waste Management,* Volume 25, No. 4, April 1982.

Levine, A. "Municipal Bond Financing for Renewable Energy Systems," Solar Energy Research Institute, Golden Colorado, 1981.

New York State Department of Environmental Conservation. *Resource Recovery Procurement and Financing,* September 1980.

Randol, R. *Resource Recovery Plant Implementation: Guides For Municipal Officials—Financing.* U.S. Environmental Protection Agency, Washington, D.C., 1976.

Shubnell, L. D. and W. W. Cobbs, "Creative Capital Financing: A Primer for State and Local Governments," *Resources in Review,* May 1982.

U.S. Conference of Mayors, *Proceedings: Resource Recovery Financing Conference,* Washington, D.C., March 1982.

Chapter 8

Freeman, H. "Pollutants From Waste-to-Energy Conversion Systems." *Environment Science and Technology,* Vol. 12, No. 12, November, 1978.

Gershman, H. W. "An Approach to Determining the Economic Feasibility of Refuse-Derived Fuel and Materials Recovery Processing." *1976 National Waste Processing Conference.* American Society of Mechanical Engineers, Boston, MA, May 1976.

Peterson, C. W. (co-author). "Drafting the Scope of Work for an Environmental Impact Statement: The Seattle Experience." *Proceedings of Third Annual Resource Recovery Conference.* National Resource Recovery Association, Washington, D.C., March 1984.

Peterson, C. W. (co-author). "Seattle's Approach to Scoping the EIS for Resource Recovery." *City Currents,* August 1984.

Peterson, C. W. (co-author). "Environmental Issues for Municipal Solid Waste to Energy Systems—Water Emissions and Residue Disposal." *Environmental Analyst,* January 1984.

Peterson, C. W. "Environmental Issues for Municipal Solid Waste to Energy Systems—Air and Emissions." *Environmental Analyst,* December 1983.

Chapter 9

Gershman, H. W. "The Planning and Development Process for Resource Recovery Projects." Paper delivered at the U.S. Conference of Mayors' Resource Recovery *Mayors Leadership Institute,* Boston, Massachusetts, July 1984.

Gershman, H. W. "The Southwest Facility," paper presented at the *Municipal Resource Recovery Conference,* Energy Bureau, Arlington, Virginia, October 1984.

Bernheisel, J. F. *Model Contract for a Full-Service Resource Recovery Program,* prepared for the U.S. Department of Energy under Contract ES-76-C-01-3851, 1980.

Barnett, T. M. and N. Kole "Get Together for Resource Recovery." *American City and County,* Volume 95, No. 11, November 1980.

Gershman, H. W. and T. M. Barnett. "How to Buy Resource Recovery." *American City and County,* Vol. 94, No. 4, April 1979.

Gershman, H. W. and T. M. Barnett. "Potholes in the Road to Resource Recovery." *Waste Age,* Vol. 9, No. 3, March 1978.

Hawkins, D. Resource Recovery Plant Implementation: Guides for Municipal Officials—Further Assistance. U.S. Environmental Protection Agency, Washington, D.C., 1976.

Lowe, R. A. and A. Shilepsky. "Resource Recovery Plant Implementation: Guides For Municipal Officials—Planning And Overview." U.S. Environmental Protection Agency, Washington, D.C., 1976.

Schoenhofer, R. F., M. A. Gagliardo and H. W. Gershman. "Fast Track Implementation of the Southwest Resource Recovery Facility, Baltimore, Maryland," In *Proceedings: 1982 National Waste Processing Conference,* New York, N.Y., May 1982.

Shilepsky, A. "Resource Recovery Plant Implementation: Guides For Municipal Officials—Procurement." U.S. Environmental Protection Agency, Washington, D.C., 1976.

Alter, H. "Resource Recovery: The Management of Technical and Economic Risks." *NCRR Bulletin,* Vol. 9, No. 4, December 1979.

Abert, J. G. "Municipal Solid Waste Recovery: A Public or Private Risk?" *Energy,* 2(2):24–26, Spring 1977.

Clunie, J. F. and M. E. Kirkpatrick. "A Review of Changes in Risk Sharing in the Financing, Construction and Operation of Solid Waste Resource Recovery Facilities." In Proceedings: 1982 National Waste Processing Conference, New York, N.Y., May 1982.

Gershman, Brickner & Bratton, Inc., "Report on the Waste-to-Energy Systems Institutional

Barriers Assessment Workshop." ANL/CNSV-TM-78, U.S. Department of Energy/ Argonne National Laboratory, Argonne, Illinois, June 1981.

Gordian Associates, Inc., "Overcoming Institutional Barriers to Solid Waste Utilization as an Energy Source." HCP/L-50172-01, U.S. Department of Energy, November 1977.

Randol, R. "Resource Recovery Plant Implementation: Guides For Municipal Officials—Risks and Contracts." Washington, D.C., U.S. Environmental Protection Agency, 1976.

Chapter 10

Gershman, H. W. "Resource Recovery and Codisposal in Auburn, Maine," in *Proceedings: 1980 National Processing Conference,* ASME Washington, D.C., May 1980.

Brickner, R. H. "Case Studies: Four Small-Scale Refuse to Energy Plants." *21st Annual International Seminar.* GRCDA, Winnipeg, Manitoba, September 1983.

Brickner, R. H. (co-author). "Case Study Review: Small-Scale Waste-to-Energy District Heating Plants." *Small-Scale Combustion Systems: A Waste-to-Energy Workshop.* U.S. Department of Energy, Argonne National Laboratory, et al., Nashville, TN, July 1983.

INDEX

American Society for Testing and Materials (ASTM), 37
Accelerated Cost Recovery System (ACRS), 146, 147, 156, 159
Acid gas scrubbers, 56, 108, 180
Aeration, 52
Air
　classifier, 76
　combustion, 58, 59
　controlled, 67
　cooling, 58
　injectors, 52
　plenums, 52
　starved, 68
　tramp, 91
　underfire (or undergrate), 50
Air pollution control equipment, 55, 56, 58, 62, 64, 68, 69, 80, 91, 108, 170
Akron Recycle Energy Systems (RES), 89
Allegheny International, 72
"All electric" projects, 127
Aluminum, 38, 90
Ancillary equipment, 84
Asbestos, 178
Ash, 186
　bottom, 55, 80
　disposal, landfill, 1
　fly, 75, 80, 91, 188
　quench-type disposal system, 52
　wet, dry, 55
Asset depreciation range (ADR), 158
Athanor, 72
Attemperators, 52
Auburn, Maine Resource Recovery Facility, 217
Avoided costs, 33, 133

Babcock & Wilcox (Madison, WI), 232–240
Back-end separation, 49
Baghouse, 56, 69
Baltimore Gas & Electric Co., 87
Basic Environmental Engineering, Inc., 72

Collegeville, MN, 72
Prudhoe Bay, Alaska, 72
Basis of Negotiations (BON), 206, 214
Belgium, 66
Beryllium (Be), 174, 178
Best Available Control Technology (BACT), 177
Best Available Demonstrated Control Technology (BADCT), 183
Black Clawson, 86
Biomass, 162
Blowback, 91
Boilers, 25–27
　dedicated, 80
　makeup water, 55
　pressure, 83
　spreader-stoker, 38
　suspension-fired, 38
　water, 50
Bond Buyers Index (BBI), 137
Bonds, 108
　adjustable rate (ARB), 152, 153–154
　daily adjustable tax-exempt securities (DATES), 152, 154
　general obligation (GO), 139–141
　industrial development revenue (IDB), 142
　municipal revenue, 141–142
　pollution control revenue, 140
　tax exempt, revenue, 108, 137–138
　variable rate demand (VRDB), 152–153
Bottom line costs, 124
Boxboard (wastepaper, corrugated, and old newspapers), 42
Bruun & Sorensen (B&S), 66
Burnout zone, 52
Business Energy Tax Credit (BETC), 162

Cadoux, S.A., 72
　Cleburne, TX, 72
　Delaware Co., PA, 72
Capital costs, 93, 106, 108–110

Carbon monoxide (CO), 173
Chlorine (Cl₂), 74
City Currents, 31
Clark-Kenith, Inc. (CKI), 64
  Alexandria/Arlington Co., VA, 64
  Hampton, VA, 64
  NH/VT Project (Claremont, NH area),
    64
  New Hanover Co., NC, 64
Clean Air Act (CAA), 59, 61, 69, 175, 178
Clean Water Act of 1977, 183, 184
Clean Air, Inc., 72–73
Coal, 6, 77, 106
  forecast, 9
  price, 8
Co-firing, 77–80, 87
Cogeneration, 6, 33, 36, 61, 63, 130
Conventional energy, 5
  current costs, 5
  future costs, 5
Combustion Engineering (CE), 87, 89
  Detroit, MI, 89
  Hartford, CT, 89
  Honolulu, HI, 89
  San Francisco, CA, 89
Combustion zone, 52
Commitment document, 48
Comtro, 72
  Jacksonville Naval Air Station, 73
Connecticut Resources Recovery Authority
    (CRRA), 13
Consumat Systems, Inc., 73
  Auburn, ME, 217–231
  Tuscaloosa, AL, 73
Corrosion, 74–75, 81
Cooling air, 50
Copper precipitation and detinning markets,
    38
Crane and Company, 241
Crane, C.P., Power Plant, 87
Credit backing, 166
Crude Oil Windfall Profit Tax Act of 1986,
    143
Cullet (glass), 40

Deaerator tank, 55
Debt service payment, 93
Deferred equity, 150–151
Deficit Reduction Act of 1984, 146, 156,
    159
Denmark, 66

Densified RDF (d-RDF), 77
Detroit, MI, 87
Detroit Edison, 87
Detroit Stoker Co. (grates), 58
Dioxin (PCDD, TCDD), 87, 174
Direct Combustion, 49, 50, 56
  Europe, Asia, 50
Dry RDF, 84
Drying zone, 52
Disposal costs, 12
Dust, 90–91
Dust RDF, 77

Economic Recovery Tax Act of 1981
    (ERTA), 159
Economics, 93, 112
Economizer, 55
Effluent, 55
Electricity,
  Asia, 60
  costs, 6, 9, 106, 112
  Europe, 60
  forecast, 9
  Gallatin, TN, 60
  generation, 23, 62
  Haverhill/Lawrence, MA, 32
  New Hanover, Co., NC, 60
Electrostatic granular filter, 179
Electrostatic precipitator (ESP), 56, 58, 69,
    178
Enercon Systems, Inc., 74
Energy answers, 88
  Albany, NY, 88
  Hamilton, Ontario, 88
  SEMASS, 88
Energy tax credit (ETC), 158, 161
Environmental impacts, 11, 105, 107, 171
Equipment, 109
European facilities, 29, 60, 84, 137
Explosions, 90
Extraction procedure (EP), 188

Fabric filters, 179
Fans
  induced draft (ID), 56, 58, 83, 91
  forced draft (FD), 52, 58, 67, 68
  overfire air fan (OFA), 83
Feasibility study, 167, 193
Federal Energy Regulatory Commission
    (FERC), 32, 33, 34

Federal Water Pollution Control Act of
    1956, 183
Ferrous Metals, 38, 76, 84
Flail mill, 76
Florida Power & Light Company, 87
Fenwel Explosion Suppression Systems, 90
Fluff RDF, 75–77
"Force Majeure," 141
Foster wheeler boilers, 58
Foundry industry, 39
Fossil fuel
    prices, 133
Fuel oil, 5
    forecast, 7
    price, 6
Fulton Iron Works, 66

General Electric, 89
General Service Contract Rule, 157–158
Glass, 40, 91
Glassphalt, 40
Grapple, 50, 56
Gravity, 50, 56, 81

Halogenated hydrocarbons, 74, 91
Hamilton-Wentworth, 89
Hauling,
    costs, 12–13, 94–103
    direct haul, 99
    time, 13
    transfer hauling, 12–13
Heating
    district, 24
    hot water district, 28–29
    space, 24
Herman Miller, Inc., 70
Hopper/chute assembly, 50, 56
Hydraulic rams, 51
Hydrogen chloride (HC1), 74, 173
Hydrogen fluoride (HF), 74, 173

Incineration, 4
Insurance, 108, 138, 168–169
    efficacy, 169
Internal Revenue Code of 1954, 140
Interest rates, 112, 137, 145, 153
Investment tax credit (ITC), 155, 158, 160–
    161
Investors, 137
Itoh. C. & Co., 65

James Madison University, 62
Japan, 60, 65, 137

Katy-Seghers, 66
    Davis Co., UT, 66
    Warren/Washington Co., NY, 66
    Savannah, GA, 66
K-Fuel, 37
Koppers Industries, Forest Products Divi-
    sion, 70
Kuhr Technologies, 89
    Saco/Biddeford, ME, 89

Landfill, 4, 49, 50, 167
    costs, 11, 17, 97, 108, 126
    environmental problems, 11
    location, 99
Lead (Pb), 173
Leveraged leasing, 146, 147–150, 151
Life cycle costs, 106, 127–130
Louvers, 52
Low pressure steam systems, 127
Lowest achievable emission rate (LAER),
    175

Madison, WI, 232
Magnetic separation, 76
Maintenance, 55, 74, 90, 108
Markets
    criteria to assess markets, 24, 44, 43–48
    waste-to-energy, 5
Maryland Environmental Service (MES), 14
Mason-Dixon Resource Developers, 89
    Henrico Co., VA, 89
Mass burning, 3, 38, 49, 55, 83
McMullen Holdings, Inc., 73
Memoranda of Understanding (MOU), 45
Mercury (Hg), 174, 178
Methane gas, 11
Modular combustion, 3, 49, 67
    Auburn, ME, 45, 70, 217–231
    Cattaraugus Co., NY, 70, 73
    Characteristics, 69
    Collegeville, MN, 70, 72
    Genesee Township, MI, 44
    North Little Rock, AR, 45, 70
    Park Co., MT, 70
    Red Wing, MN, 70
    Siloam Springs, AR, 69
    Tuscaloosa, AL, 70, 73

Modular combustion (*cont.*)
  under construction
    Alameda, CA, 70
    Durham, NH, 70
    Oneida Co., NY, 70
    Oswego Co., NY, 70
    Portland, ME, 70
    Rutland, VT, 70
    Springfield, MA, 70
    vendors, 70–74
    Waxahachie, TX, 70, 73
    Windham, CT, 45
    Zeeland, MI, 70
Montenay International Corporation of
    NY, 67
Moody's Investors Services, 170
Morse-Boulger, Inc. (M-B), 67
  Glen Cove, NY, 67
  Harrisonburg, VA, 67
  Hempstead, NY, 67
  "Kascade" Stoker, 67
Mustang RDF Co., 89
Municipal solid waste (MSW), 4, 50, 59, 60,
    126, 171
  as a fuel, 5
  composition, 17
  crisis, 193
  energy content, 19, 76
  for cogeneration, 6
  seasonal variation, 17

Natural gas, 4, 6, 7, 106
Natural Gas Policy Act (NGPA), 8
National Center for Resource Recovery
    (NCRR), 39
National Ecology, Inc., 89
  Baltimore Co., MD, 89
National Ambient Air Quality Standards
    (NAAQS), 172, 175, 190
National Emission Standards for Hazardous
    Air Pollutants (NESHAP), 174, 175,
    178
National Pollutant Discharge Elimination
    System (NPDES), 183
National Pretreatment Standards, 184, 186
New Source Performance Standards
    (NSPS), 75, 178
New York State Department of Environ-
    mental Conservation, 165
New York State Environmental Facilities
    Corporation, 14

Nitric oxides (NO$_X$), 172
Nonmagnetic metals (nonferrous), 41
  Dade Co., FL, 71, 72
Northeast Maryland Waste Disposal Au-
    thority, 13
Nuclear energy, 106

O'Connor Combustor Corporation, 51, 54,
    65
  Bay Co., FL, 65
  Dutchess Co., NY, 65
  Gallatin, TN, 65
  Los Angeles, CA, 65
  West Contra Costa Co., CA, 65
Odor, 77, 87
Ogden-Martin (O-M) Corporation, 65
  Chicago, IL, 65
  Harrisburg, PA, 65
  Pinellas Co., FL, 65
  Portland, ME, 65
  Tulsa, OK, 65
Oil
  imported, 5
  prices, 4, 6
Operation and maintenance costs (O&M),
    93, 96–97, 108, 111–169
Organic-Fuel 100™, 37
Organization of Petroleum Exporting
    Countries (OPEC), 3, 6, 87
Oscar Mayer & Co., 236
Overhead bridge crane, 50
Oxygen, 55, 67

Payments in lieu of taxes (PILOTs), 124
PEDCo, 26
Parsons & Whittemore, 87, 88
  Dade Co., FL, 87
  Hempstead, NY, 87
Particulates, 172
Pellets, 77
Pennsylvania Engineering, Inc., 65
Penobscot Energy Recovery Company, 89
  Bangor, ME, 89
Permits, 109
Petroleum
  prices, 5
  unregulated, domestic, 5–6
Pig iron scrap, 39
Piping systems, 29
Pittsfield, MA, 241
Plastics, 18, 49, 74, 173

Pioneer Plastics, 218
Post consumer waste, 40
Prevention of significant deterioration
    (PSD), 175, 177
Processed refuse fuel (PRF), 37
Procurement, 203
Proposition 13 (California), 138
Proposition, 2 $\frac{1}{2}$ (Massachusetts), 138
Publicly owned treatment works (POTW),
    184
Public involvement, 215
Public Utility Holding Company Act, 33
Public Utility Regulatory Policy Act
    (PURPA), 6, 10, 33, 34, 133, 136
Pyrolysis, 3, 82

Refuse-derived fuel (RDF), 3, 37, 38, 40,
    75–88, 127, 171, 182, 187
  Albany, NY, 86
  Akron, OH, 86
  Ames, IA, 38, 85
  characteristics, 84
  co-firing, 77–80
  Columbus, OH, 87
  cost, 80, 83
  Dade County, FL, 40, 87, 92
  dry, 75–77
  Hamilton, Ontario, 86
  Haverhill/Lawrence, MA, 87
  Hempstead, NY, 87
  Lakeland, FL, 85, 86
  Madison, WI, 38, 84–85, 332
  Niagara Falls, NY, 83, 86
  St. Louis, MO, 85
  vendors, 88–89
  wet, 77
  Wilmington, DE, 40
Resource Conservation & Recovery Act
    (RCRA), 11, 96, 187
RECO Industries, Inc., 73
Recycling, 40, 41
  aluminum, 41
  glass, 40
  paper, 42
  plastic, 173
Refractory-lined incinerators, 50, 56,
    60–63, 66
  efficiency, 58
  Glen Cove, NY, 62, 63
  Harrisonburg, VA, 62, 63
  Hamburg, West Germany, 60

New York, NY, 60
  Betts Avenue Incinerator, 62
  under construction
  Davis Co., UT, 63
  Key West, FL, 63
  Savannah, GA, 63
  Susanville, CA, 63
  Tampa, FL, 63
  Warren/Washington Co., NY, 63
  Waukesha, WI, 58, 62
  vendors, 66
Refuse Fuels, Inc., 87
Request for Developers (RFD), 214
Request for Proposals (RFP), 204
Request for Qualifications (RFQ), 206
Residue, 186
Revenue, 5, 93, 126, 167
Revenue recovery, 36
Riley Stoker Corporation, 65

Salt Lake City, UT, 45, 46
Scrubbers (wet, dry), 180
Sigfi, SA, 73
Signal-RESCO (S-R), 52, 65
  Baltimore, MD, 65
  North Andover, MA, 65
  Pinellas Co., FL, 65
  Saugus, MA, 65
  Westchester Co., NY, 65
Sigoure Freres, S.A., 73
  St. Lawrence Co., NY, 74
  Sitka, AK, 74
Silicon carbide (SiC), 74, 92
Site development, 109
Sludge, 30
Small power production facility, 33
Source separation, 40
Sparklers, 52
Standard & Poor's Corporation, 170
Standard Industrial Code (SIC), 27
Standard Metropolitan Statistical Area
    (SMSA), 39
Steam, 28, 83
  contract, 168
Steel industry, 38
Step financing, 151
St. Joseph's Hospital (Hot Springs, AR), 69
Stoichiometric condition, 52, 68, 83, 172
Storage pit, 50
Sulfur oxides (SO_x), 173
Sunbeam Corp., 72

Superheater, 52, 58, 65, 83
Synergy Systems Corporation, 73
Synergy-Clear Air, 73
  Cattaraugus Co., NY, 73
  Dade Co., FL International Airport, 73
  Waxahachie, TX, 73

Takuma Ltd., 60, 65
Tampa Electric Co., 63
Taylor, R.W. Group/Clear Air, Inc., 73
  Cattaraugus Co., NY, 73
  Ogden, UT, 73
  Oneida Co., NY, 73
Tax credit, 161
Tax Equity and Fiscal Responsibility Act of
    1982 (TEFRA), 146, 156, 161
Tax-exempt bonds, 5
Teledyne National, 89
  Akron, OH, 89
Temperature, 68–69
Thermal transfer fluids, 22–30
300 Fuel™, 37
Tin, 39
Tipping fee, 1, 93, 136
Total hydrocarbons (THC), 173
Transfer station, 99, 100
  costs, 101
Transportation, 97
  grid, 100
  of RDF, 127
  routes, 98
  vehicle impacts, 97
Tricil, Ltd., 89
  Akron, OH, 89
  Hamilton, Ontario, 89
Trommel screen, 76, 91
Turbine-generator (T-G) set, 131

U.S. Department of Energy (DOE), 7, 27,
    36, 89
U.S. Department of Housing & Urban De-
    velopment (HUD), 29
U.S. Department of the Interior, Bureau of
    Mines, 44
U.S. Environmental Protection Agency
    (EPA), 11, 16, 18, 26, 44, 85, 86, 87,
    174, 183

U.S. Treasury Department, 143
  Internal Revenue Service (IRS), 147, 156
Urban Development Action Grant (UDAG),
    29
Utility grid, 109

Vicon Recovery Associates, 74, 241–257
  Pittsfield, MA, 74, 241–257
  Wilmington, DE, 74
Vinyl chloride, 178
Volund USA (VUSA), 66
Von Roll technology, 65

Wallop Amendment, 157
Waste heat boiler, 58
Waste management, 66
  Tampa, FL, 66
Waste materials, combustible and noncom-
    bustible, 49
Waste sheds, 98, 100
Waste stream, 98, 99
Water
  groundwater contamination, 11
  surface runoff, 11
  wastewater, 181
Water Quality Act of 1965, 183
Waterwall incinerators, 50, 52, 64, 83
  Bern, Switzerland, 58
  efficiency, 58
  Europe, 58
  Gallatin, TN, 60
  New Hanover Co., NC, 60, 64–65
  Norfolk, VA, 58–59
  Saugus, MA, 52
  under construction
    Bay Co., FL, 63
    Commerce City, CA, 63
    Marion Co., OR, 62
    Olmstead Co., MN, 63
    Tulsa, OK, 63
  vendors, 64
Westinghouse (see O'Connor Combustor
    Corp.)
Wet RDF, 77
Wheeling (electrical), 32
White goods (bulkies), 49, 75

Zink, John Co. (see also Comtro), 72